ADVENTURES OF THE BRAIN

WRITTEN BY PROFESSOR SANJAY MANOHAR
ILLUSTRATED BY GARY BOLLER

WAYLAND

CONTENTS

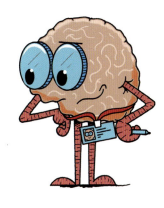

WHAT IS THE BRAIN?

The brain is a soft, jelly-like organ in your head. It is the part of your body that thinks, feels, has ideas, and performs actions.

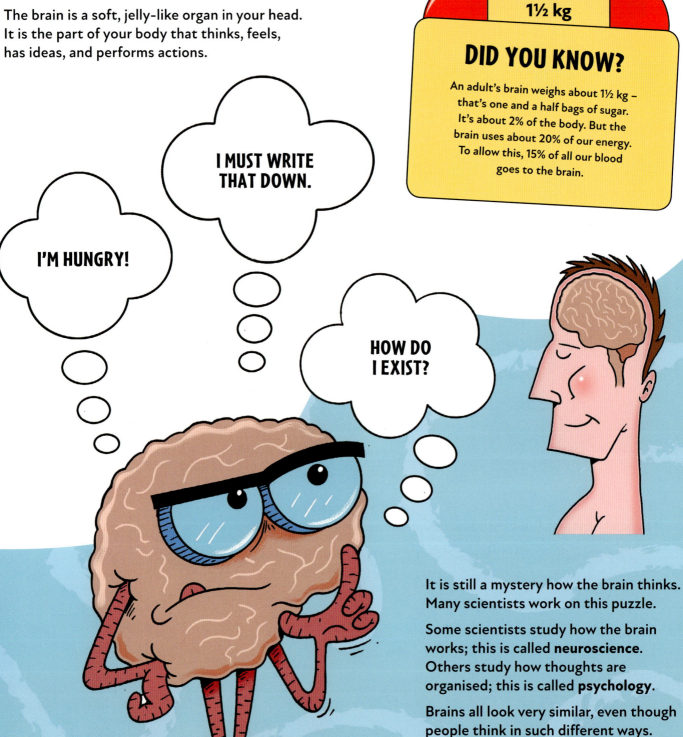

1½ kg

DID YOU KNOW?

An adult's brain weighs about 1½ kg – that's one and a half bags of sugar. It's about 2% of the body. But the brain uses about 20% of our energy. To allow this, 15% of all our blood goes to the brain.

I'M HUNGRY!

I MUST WRITE THAT DOWN.

HOW DO I EXIST?

It is still a mystery how the brain thinks. Many scientists work on this puzzle.

Some scientists study how the brain works; this is called **neuroscience**. Others study how thoughts are organised; this is called **psychology**.

Brains all look very similar, even though people think in such different ways.

The brain's job is to keep us alive. It needs oxygen from the air and energy from food, to work. These are supplied by blood:

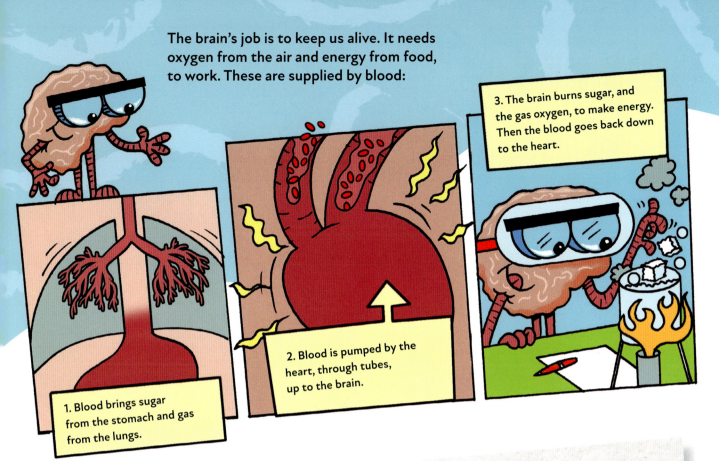

3. The brain burns sugar, and the gas oxygen, to make energy. Then the blood goes back down to the heart.

1. Blood brings sugar from the stomach and gas from the lungs.

2. Blood is pumped by the heart, through tubes, up to the brain.

PARTS OF THE BRAIN

The brain is made up of different parts. They do different things.

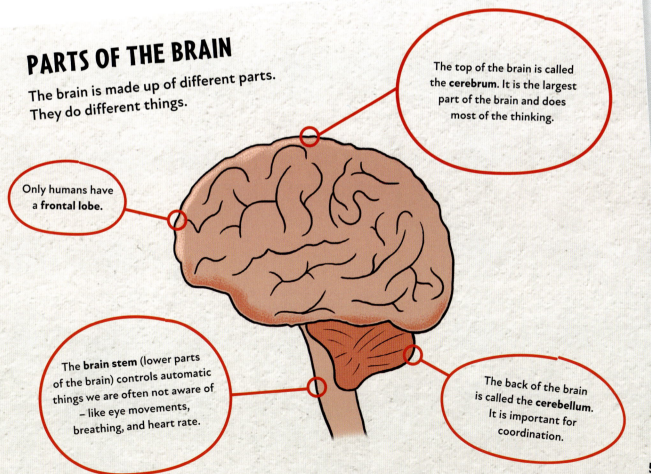

The top of the brain is called the **cerebrum**. It is the largest part of the brain and does most of the thinking.

Only humans have a **frontal lobe**.

The **brain stem** (lower parts of the brain) controls automatic things we are often not aware of – like eye movements, breathing, and heart rate.

The back of the brain is called the **cerebellum**. It is important for coordination.

5

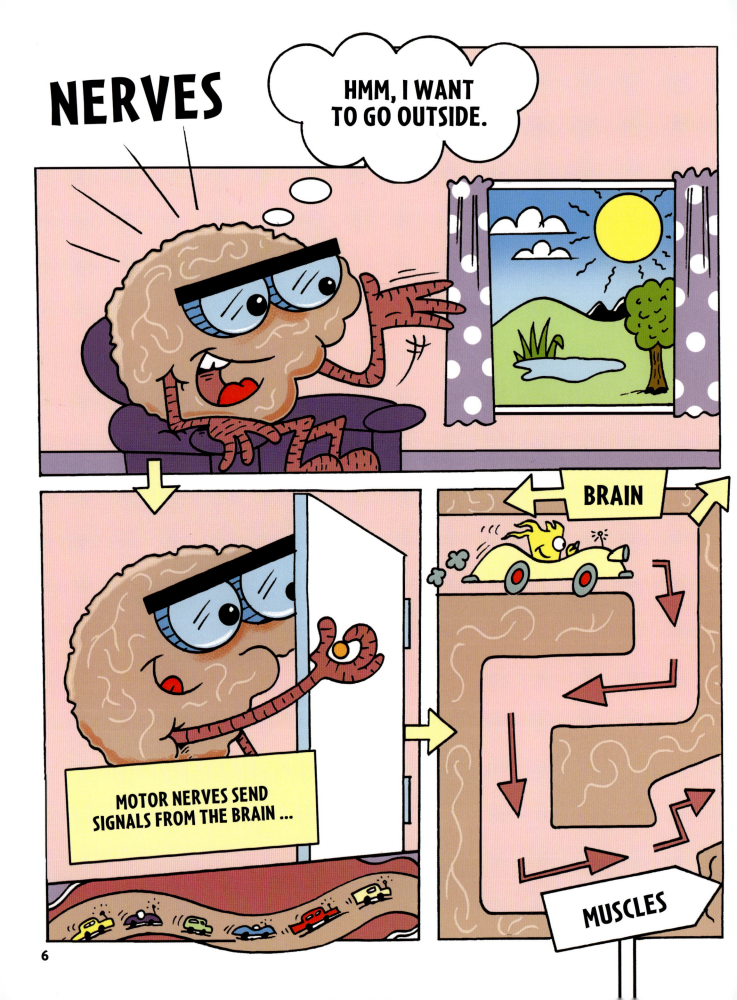

7

HOW NERVES WORK

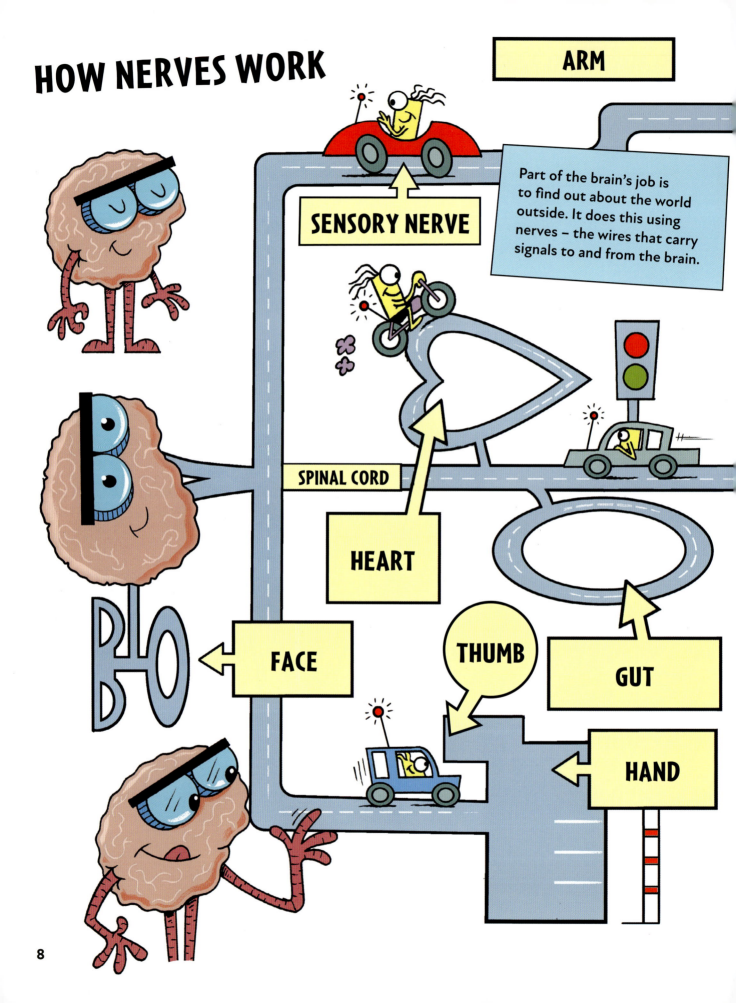

ARM

SENSORY NERVE

Part of the brain's job is to find out about the world outside. It does this using nerves – the wires that carry signals to and from the brain.

SPINAL CORD

HEART

FACE

THUMB

GUT

HAND

SLOW

LEG

MOTOR NERVE

Wires going towards the brain are called **sensory nerves.** They tell you about what is outside your body. Wires going away from the brain are called **motor nerves.** They tell muscles when to move.

The lowest part of the brain connects to the rest of your body. Wires go up and down into the **spinal cord.** This is a thin strip as wide as your little finger, that runs all down your back.

FOOT

KNEE

DID YOU KNOW?

The longest nerve in your body goes down to your big toe!

9

NEURONS

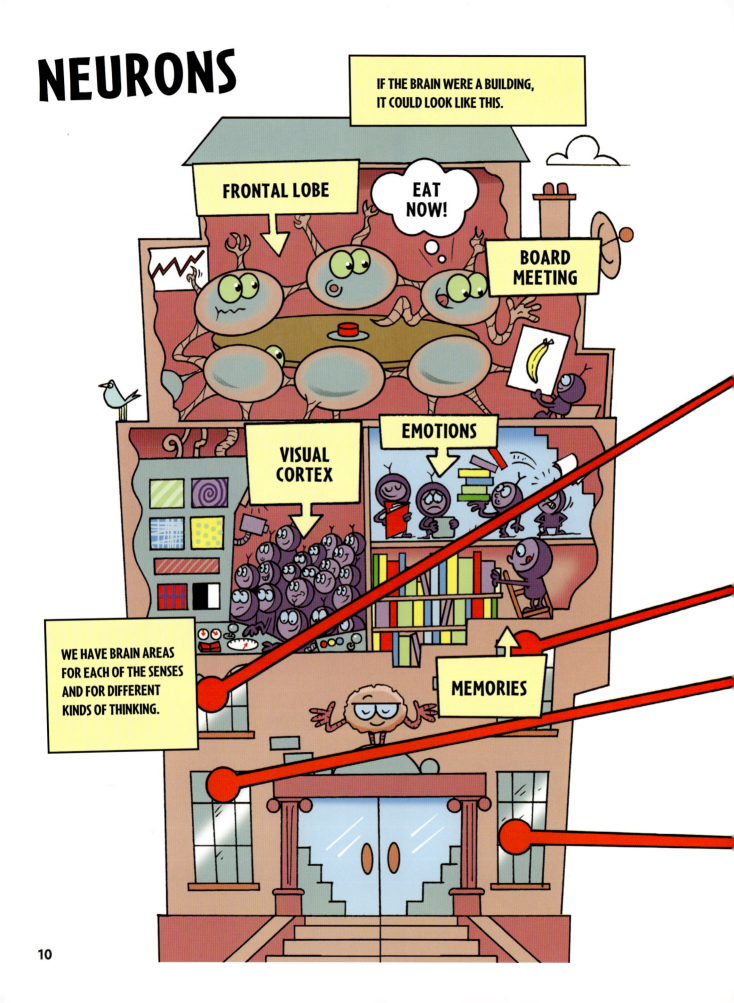

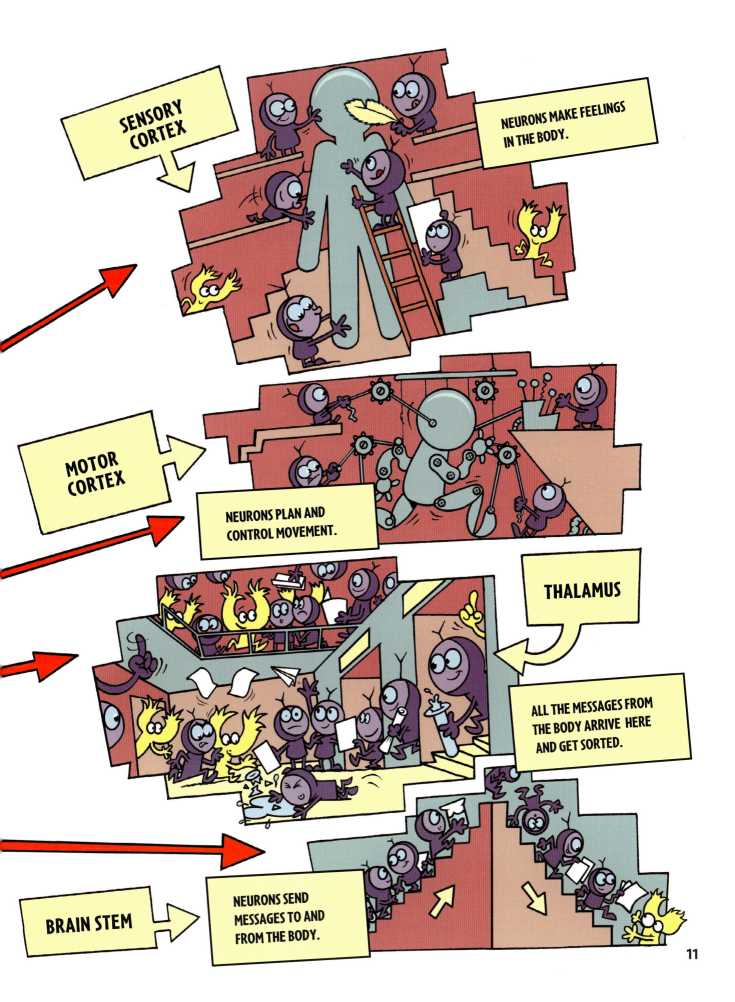

HOW NEURONS WORK

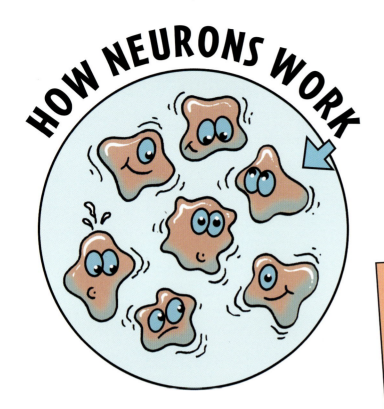

Cells are the building blocks of all living things They are very tiny! You could fit 10,000 cells inside this full stop. Cells stick together in patterns, to make different materials called tissues.

The brain is made up of special cells called neurons. We have about 100,000,000,000 (one hundred billion) neurons. Helper cells surround neurons and make sure they have all the food and oxygen they need.

HELPER CELL

THE BODY THINKS IT'S HUNGRY.

Neurons talk to each other though their long arms. One neuron has arms touching thousands of other neurons. Their fingers touch at synapses.

The neurons listen to each other, and respond in different ways. Together, they produce thoughts and control what we do.

SYNAPSE

NO IT HAS EATEN ALREADY!

DID YOU KNOW?

There are two types of tissue in the brain: white matter and grey matter. Grey matter is mainly made of the neurons themselves. White matter is all their 'arms' – the bundles of wires connecting grey matter together.

HOW MESSAGES WORK

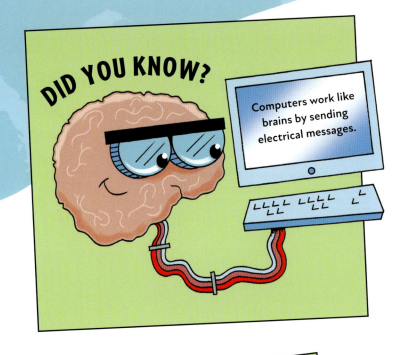

DID YOU KNOW?

Computers work like brains by sending electrical messages.

Neurons are either active or quiet. When a neuron becomes active, it sends an electric signal down its arms (called axons). Electric signals tell the end of the arm to wake up the next neuron.

When a neuron is active, its arms pass on the signal by making a chemical. The next neuron feels this chemical. If enough chemicals reach a neuron, it will also become active. It can then pass on the electrical message again. The chemicals go away very quickly, so that signals can start and stop very fast.

AXON

Signals travel down the arms very quickly. They can travel 100 metres in one second. **That means that in one second, a message can go from one end of your brain to the other 1,000 times!**

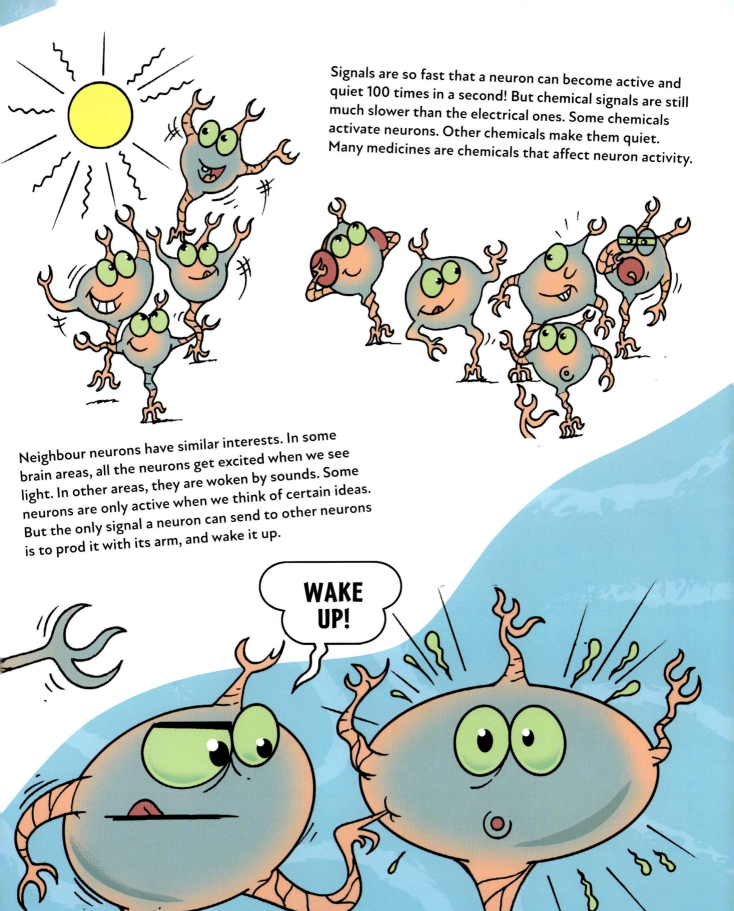

Signals are so fast that a neuron can become active and quiet 100 times in a second! But chemical signals are still much slower than the electrical ones. Some chemicals activate neurons. Other chemicals make them quiet. Many medicines are chemicals that affect neuron activity.

Neighbour neurons have similar interests. In some brain areas, all the neurons get excited when we see light. In other areas, they are woken by sounds. Some neurons are only active when we think of certain ideas. But the only signal a neuron can send to other neurons is to prod it with its arm, and wake it up.

SEEING

BRAIN LOOKS OUTSIDE AS A DOG GOES BY.

WE SEE THINGS WHEN OUR EYES DETECT PATTERNS OF LIGHT SHINING ON THEM.

LIGHT

PUPIL

SOME NEURONS BECOME ACTIVE WHEN THEY NOTICE SOMETHING CHANGE IN THE WORLD. THEY ARE CALLED RECEPTORS. AT THE BACK OF THE EYE THERE IS A SHEET OF SPECIAL NEURONS THAT ARE LIGHT RECEPTORS. THESE BECOME ACTIVE OR QUIET WHEN LIGHT SHINES ON THEM.

RECEPTOR

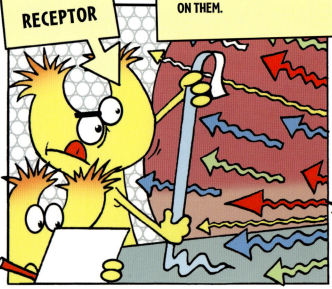

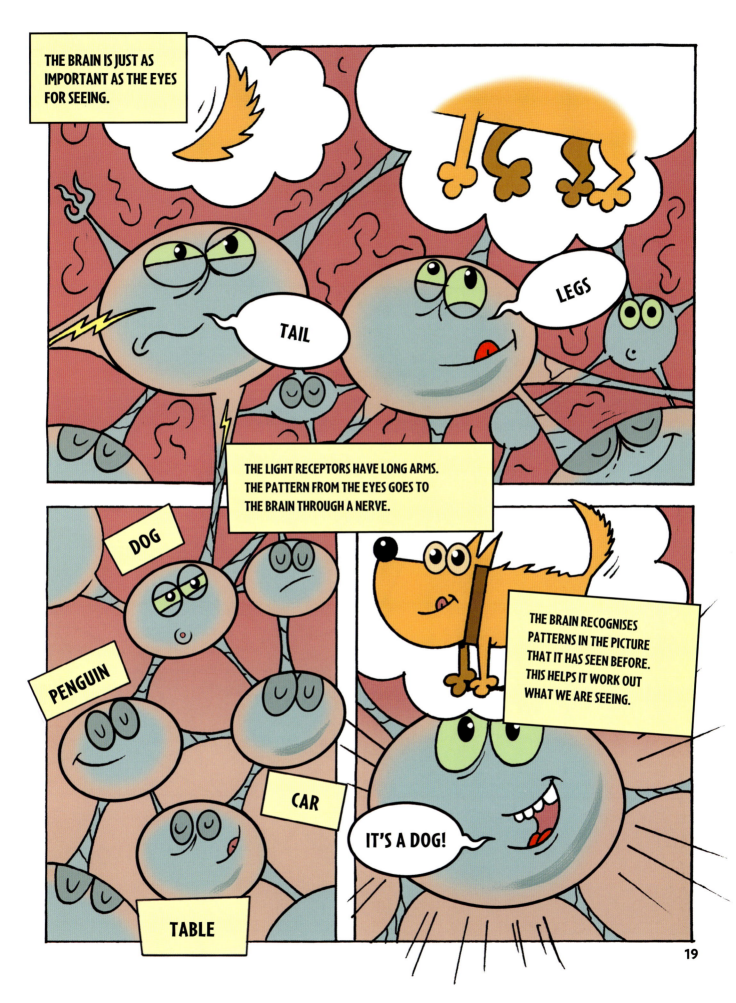

HOW SEEING WORKS

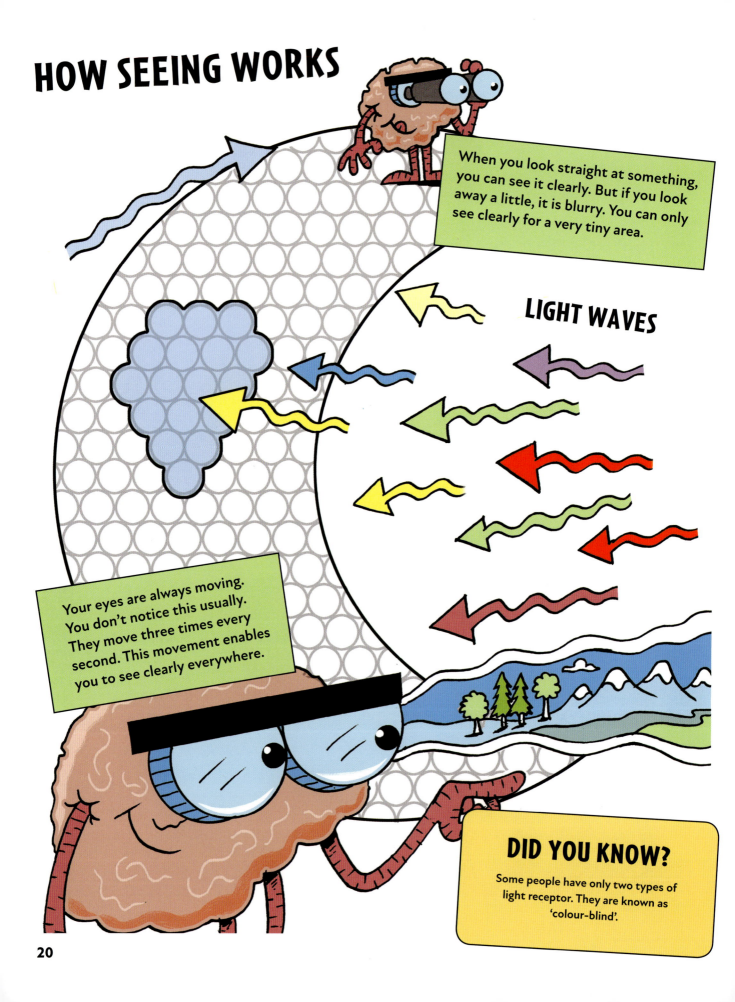

When you look straight at something, you can see it clearly. But if you look away a little, it is blurry. You can only see clearly for a very tiny area.

LIGHT WAVES

Your eyes are always moving. You don't notice this usually. They move three times every second. This movement enables you to see clearly everywhere.

DID YOU KNOW?

Some people have only two types of light receptor. They are known as 'colour-blind'.

Humans have light receptors to detect three types of light. This is what gives us colours. Because we have three receptors, we can tell apart three primary colours. With just three types of light, we can make up red, green, blue, and all the shades in between.

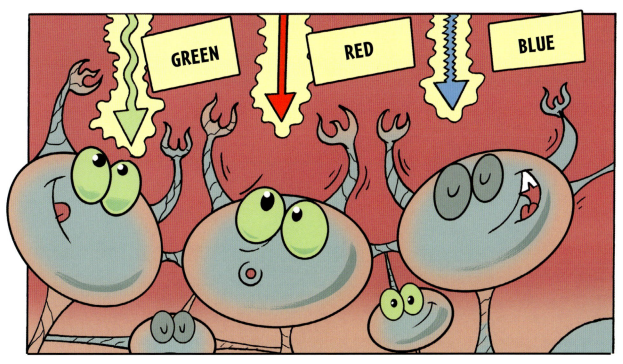

You can see in the dark, or in the bright day. When it is dark, your eyes let more light in. The pupil gets bigger. On a bright day, it shrinks. In the dark, you cannot see colours.

CONCENTRATION AND DISTRACTION

ATTENTION ACTS LIKS A DOOR BETWEEN TWO PARTS OF THE BRAIN.

BRAIN IS READING A BOOK WHILE WALKING IN THE JUNGLE.

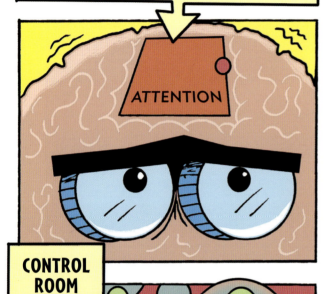

ATTENTION

CONTROL ROOM

SNAKE!

I'M SORRY WE ARE READING A BOOK.

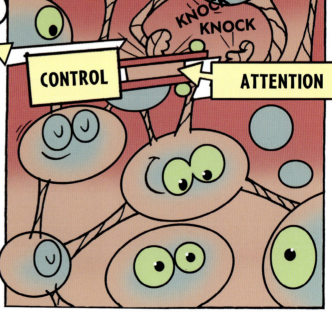

KNOCK KNOCK

CONTROL

ATTENTION

HOW CONCENTRATION AND DISTRACTION WORKS

Each of our senses is always sending signals to the brain.

We get light signals from the eyes, sound signals from the ears, as well as touch signals from all over the body.

But you can't think about all of these, all the time. There is a part of your mind called 'attention' that chooses which bits to think about. It lets you focus on one thing.

DID YOU KNOW?

Some people find it harder to focus attention than others. This is sometimes called 'attention deficit' and these people take longer to do some things.

ATTENTION DEPENDS ON TWO THINGS

1) The frontal lobe decides what you'll need to think about. Maybe it will focus on your teacher in class.

2) The receptors send urgent messages, when they want to grab attention. Maybe a loud noise behind you will make you turn around (even when you should be looking at your teacher).

Paying attention can be hard. Scientists do not yet understand why it is hard to pay attention. Maybe it is because, if you were an animal in the jungle, then paying too much attention to one thing could mean you ignore a danger.

HISSSSS!!

In a jungle, it is better to turn around if there is a loud noise behind you!

FEELING PAIN

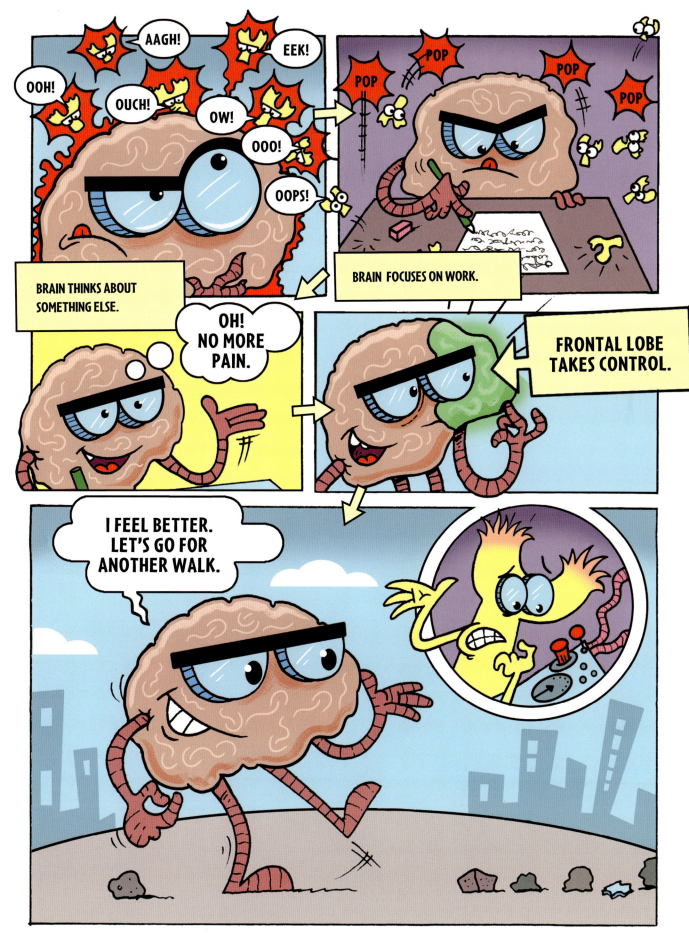

WHAT IS PAIN?

Pain is our safety mechanism. We have receptors for all types of extremes:
- Is it too hot?
- Is it too cold?
- Is something pressing too hard?
- Has my skin got cut or damaged?

Pain signals travel from your skin, through nerves, to your spinal cord then up it, to the centre of the brain. This triggers off your reactions to pain.

Receptors react to changes in our body.

When anything like this happens, the brain does the same thing ...

"OW!" Stop what you are doing and protect yourself.

This same nerve pathway is used by most animals with a backbone – fish, reptiles, birds, and mammals.

STOPPING PAIN

Touching the skin near a hurt area can block pain signals.

Pain feels stronger when we are concentrating on it. Doing other things that need brainpower can reduce pain.

Medicines called painkillers can block the nerve signals. They make pain feel less bad.

Cooling the skin down can slow down nerves that signal pain.

PAIN CAUSES FEAR

Fear can be useful – it makes animals stay away from harm. But for people, fear can worsen pain. **So be brave!**

DID YOU KNOW?

Some people meditate. This means they learn to control the focus of their mind. Meditation may block pain signals from entering the brain, just by the power of thinking.

WHAT DOES THE BRAIN DO?

The brain is always busy. It has a lot of jobs:

ACTION

SEEING

REMEMBERING

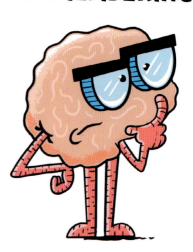

HEARING

FEELING

I WANT TO GO FOR A WALK.

DECIDING

Do we see with our eyes or our brain? Eyes detect light. But the brain interprets the light pattern and gives it meaning.

HAVE I SEEN THIS PATTERN BEFORE?

APPLE

AHA! A MATCH

WHAT DOES IT MEAN? CAN I EAT IT? CAN I SIT ON IT?

It also knows these answers, from memory.

How does it do this? It has a memory of things we have seen and compares them all.

DID YOU KNOW?

Babies need to learn to see. Once their brains have enough memories, they can start to recognise things. It takes weeks for babies to recognise faces or follow things that move.

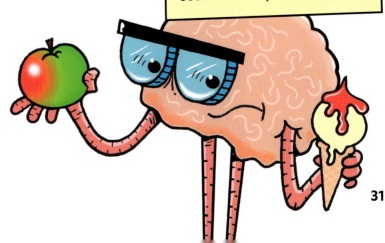

31

BRAIN VERSUS COMPUTER

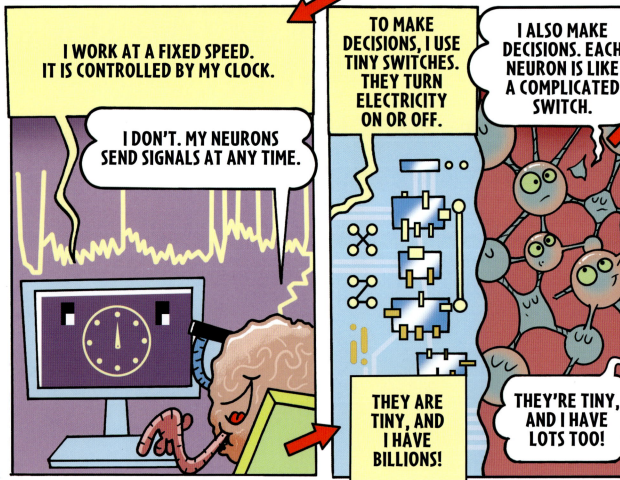

I CAN SAVE THINGS IN MY MEMORY. I HAVE SPECIAL SWITCHES TO KEEP MEMORIES.

I ALSO KEEP SOME THINGS IN MY MEMORY. I HAVE SPECIAL NEURONS FOR THAT.

I HAVE A CAMERA – WHEN LIGHT HITS IT, I GET ELECTRICAL SIGNALS DOWN THIS WIRE.

MY EYE DOES THAT. THE ELECTRICAL SIGNAL COMES DOWN A NERVE.

WIRE

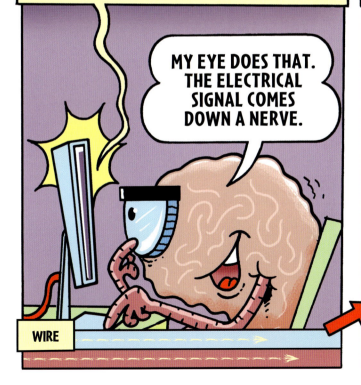

WHEN YOU PRESS A KEY, I COMPARE IT WITH MY MEMORY. THEN I MAKE A DECISION ABOUT WHAT TO DO.

WHENEVER I SEE ANYTHING, I COMPARE IT WITH MY MEMORY. THEN I MAKE A DECISION ABOUT WHAT TO DO.

HOW A BRAIN COMPARES TO A COMPUTER

Both brains and computers decide what to do.
But they work differently.

Computers are made of wires and silicon switches, but the brain is made of living cells. Computers send signals by making the electricity level high or low, along a wire. They can do this thousands of millions of times a second.

Brain cells send pulses of electricity down their arms – but only a hundred times a second.

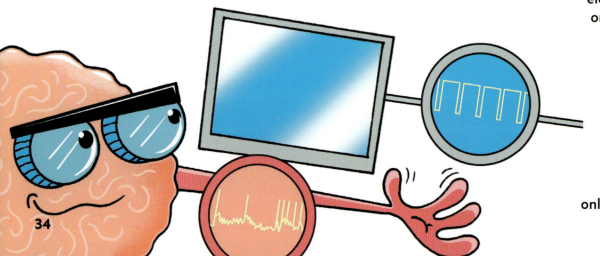

A computer might have 10 billion decision points on it.

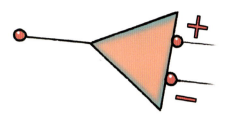

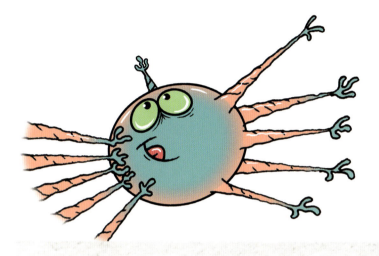

The human brain may have about 100 billion cells. Each neuron acts as a decision point, and gets signals from at least 1,000 other neurons, whereas decision points in a chip only get 2 signals.

BRAIN VS COMPUTER

Computers are designed by humans. Everything they do is "programmed": they follows rules made by people.

Our brains are not. Most things we do are learned by watching and trying things out.

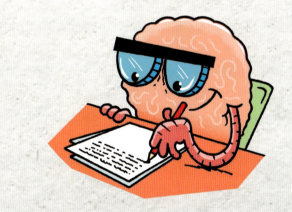

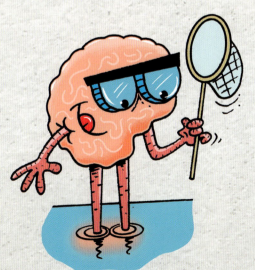

REFLEXES

CRASH!

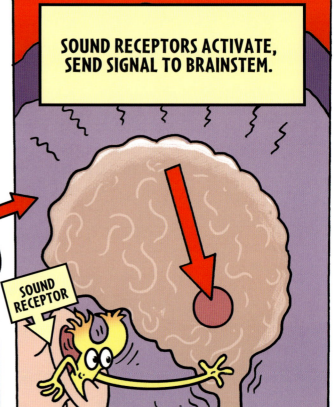

SOUND RECEPTORS ACTIVATE, SEND SIGNAL TO BRAINSTEM.

SOUND RECEPTOR

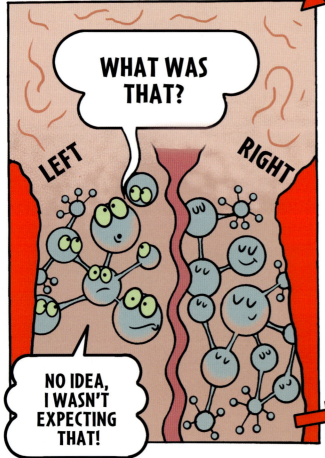

WHAT WAS THAT?

LEFT

RIGHT

NO IDEA, I WASN'T EXPECTING THAT!

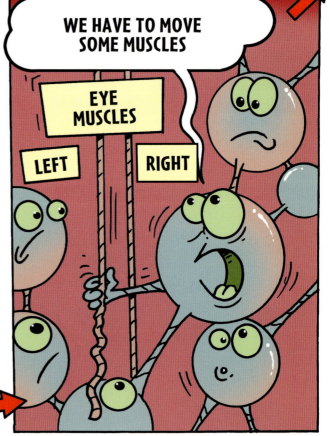

WE HAVE TO MOVE SOME MUSCLES

EYE MUSCLES

LEFT

RIGHT

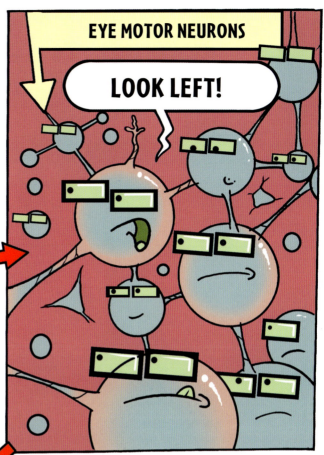

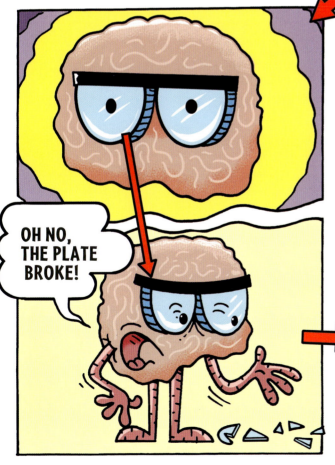

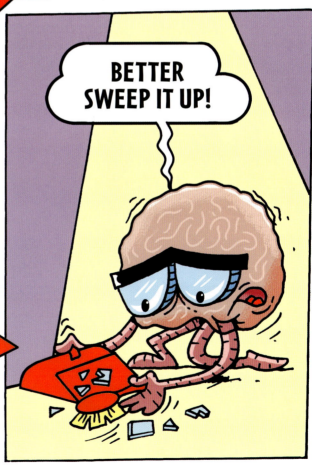

HOW REFLEXES WORK

Reflexes are the simplest responses we can make. Startle reflexes can make us look round quickly, even without thinking.

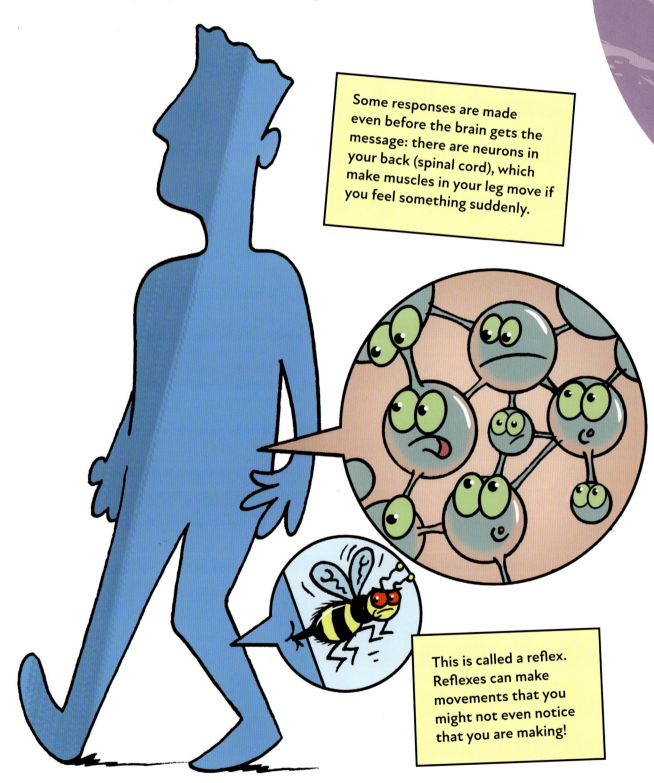

Some responses are made even before the brain gets the message: there are neurons in your back (spinal cord), which make muscles in your leg move if you feel something suddenly.

This is called a reflex. Reflexes can make movements that you might not even notice that you are making!

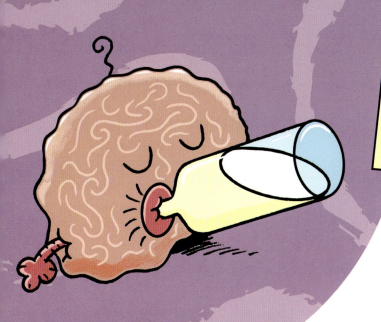

BABIES HAVE MANY SIMPLE REFLEXES THAT WORK FROM BIRTH.

THEY SUCK MILK BY REFLEX.

MANY OF THESE ARE SWITCHED OFF WHEN YOU ARE A FEW YEARS OLD.

MOTIVATION AND REWARD

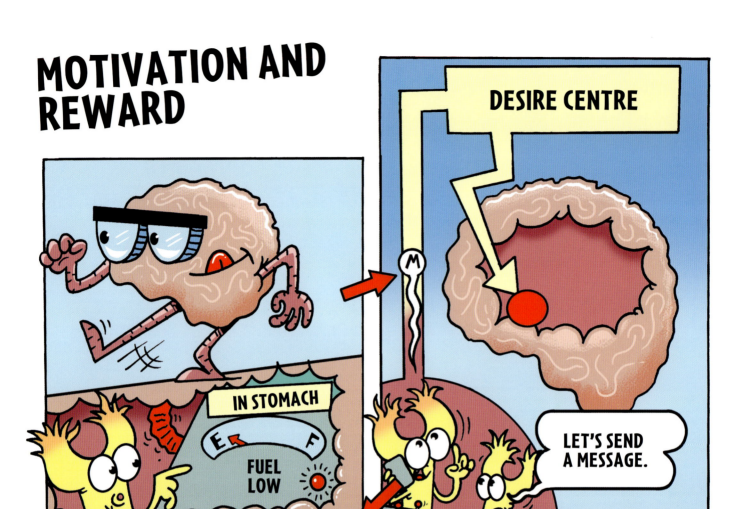

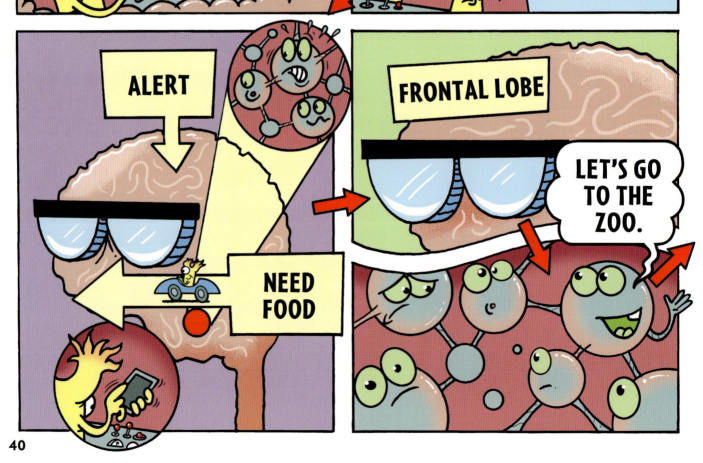

HOW MOTIVATION AND REWARD WORKS

The brain wants different things at different times. Sometimes it is hungry. Sometimes it is thirsty. Desires give us motivation, telling us what to do.

If I am thirsty, having a drink is rewarding. Rewards are nice and make us happy. Someone smiling at you is rewarding too.

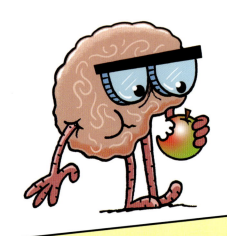

The opposite of rewards are punishments, or penalties. Penalties are things that the brain doesn't like.

Getting hurt, tasting bitter food and being told off are penalties. Missing out on a reward is also a penalty. Penalties also motivate us. We try to avoid them.

We can learn new kinds of rewards and penalties. We learn that scoring a goal, winning points, or earning money are also rewarding. Babies don't know how to get what they need. They can only cry. Rewards (like getting milk) help them learn what to do.

Learning changes our behaviour. We learn from both rewards and penalties.

Sometimes it is hard to know what caused a reward. Imagine you baked a cake, and it tasted much better than usual. You put more butter and less flour in. Was the cake better because it had more butter, or less flour? It is hard to know. But to learn anything, your brain has to solve these puzzles all the time.

People tend to repeat actions that are rewarding. But this can create an endless loop. If you get your favourite sweets every time you press a button, you would keep pressing it.

Sometimes it can be very hard to stop!

MOVING ABOUT

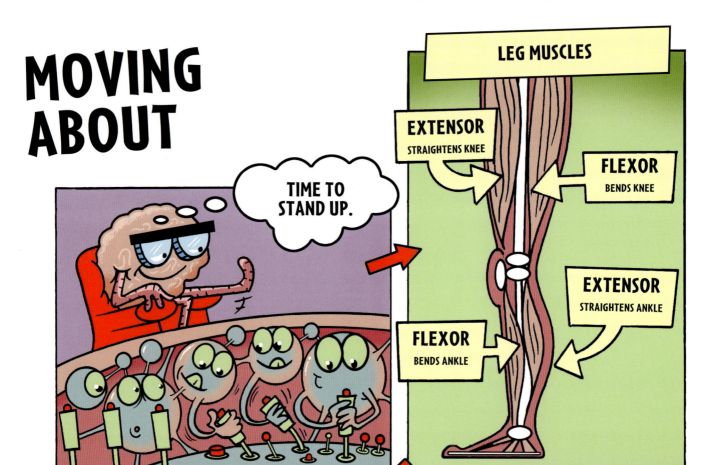

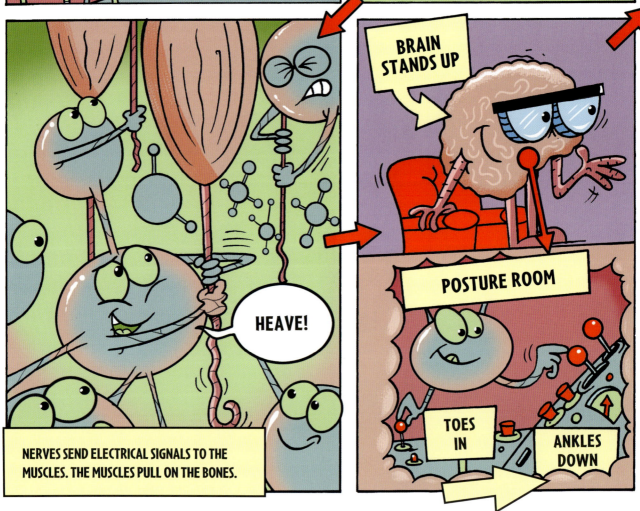

NERVES SEND ELECTRICAL SIGNALS TO THE MUSCLES. THE MUSCLES PULL ON THE BONES.

WHOA! IT'S WINDY.

IF SOMETHING PUSHES YOU, YOUR BRAIN HAS TO REACT, TO STOP YOU FALLING. THIS IS CALLED POSTURE. THE BACK PART OF THE BRAIN CONTROLS POSTURE.

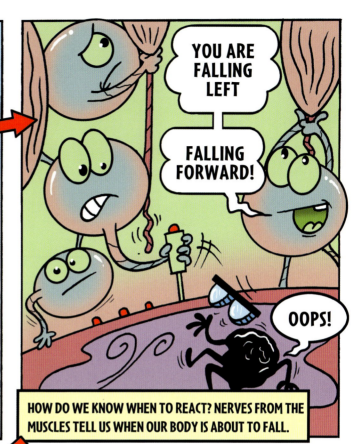

YOU ARE FALLING LEFT

FALLING FORWARD!

OOPS!

HOW DO WE KNOW WHEN TO REACT? NERVES FROM THE MUSCLES TELL US WHEN OUR BODY IS ABOUT TO FALL.

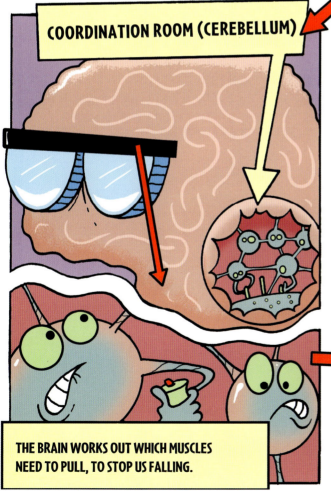

COORDINATION ROOM (CEREBELLUM)

THE BRAIN WORKS OUT WHICH MUSCLES NEED TO PULL, TO STOP US FALLING.

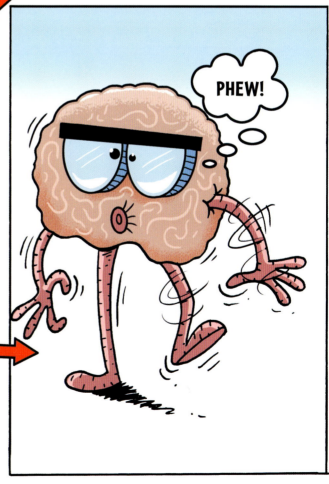

PHEW!

HOW MOVING WORKS

Muscles make movement by pulling on the body. The brain controls about 600 muscles. It controls your arms, legs, spine, hands, feet and face.

Every little movement uses different combinations of muscles. How do we know which muscles to move, and when? The brain controls each one with separate nerves.

DID YOU KNOW?

Standing up needs more than 50 muscles, all pulling just the right amount.

Standing is hard, but running and jumping are even harder!

Each muscle has to work at exactly the right time. This is called coordination. The brain does lots of maths to work out the right forces.

LONG JUMP

$$(1 + 2l) - 2(X) + 5 = 1 + 4l^2 + 3 - 2 - 4l + 8^5$$

The cerebellum at the back of the brain is mainly for coordination.

Before we move, we **decide** to move. This is when your brain thinks about what to do. When the brain sees something, like a ball to catch, it takes half a second to react.

OWZAT

The brain weighs up all the options. If we are not sure, it takes longer to react.

CONTROL

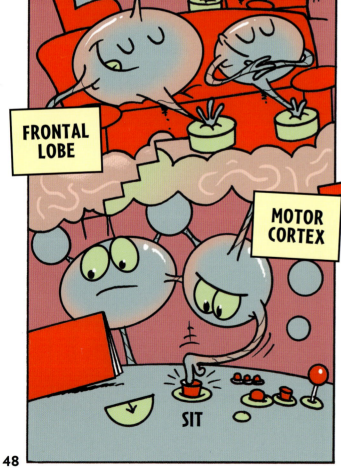

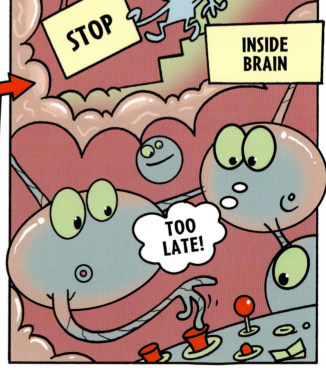

49

HOW CONTROL WORKS

The brain sometimes does things without thinking them through. We need extra control when we have to stop doing something that's automatic. It takes effort and concentration. Large parts of the brain's frontal lobe are active when we stop and take control.

The frontal lobe helps us be less automatic, but also helps us resist things we want right now – to resist an urge. Young children have less control.

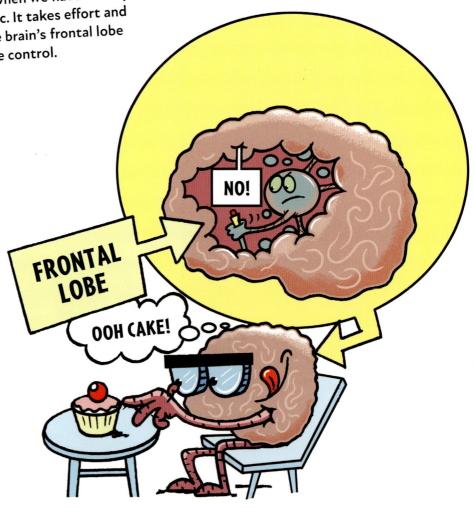

Using control, even when it's hard, is called willpower. You need willpower to eat your main meal before pudding. You need willpower to practise hard. Willpower helps you succeed.

DID YOU KNOW?

Your brain is made up of 60% fat, making it the fattest organ in the human body!

Too little control can lead to bad behaviour, like disobeying rules, or unhealthy eating. While it feels good now, it will be bad for you later.

GURGLE

RUMBLE

ACHE

PLANNING AHEAD

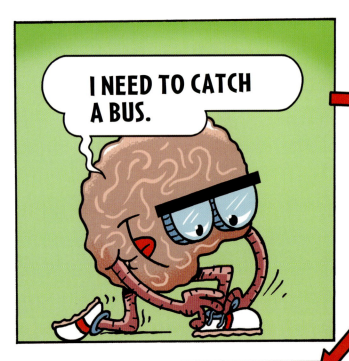

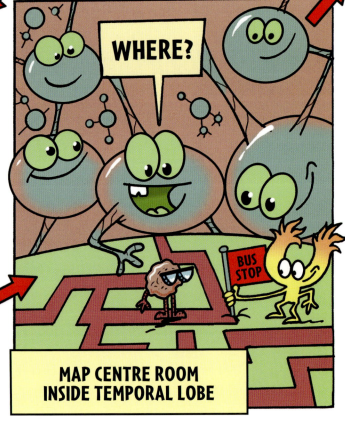

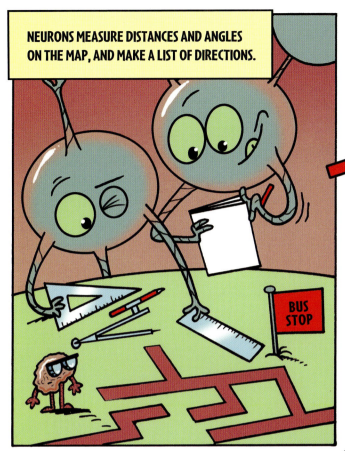

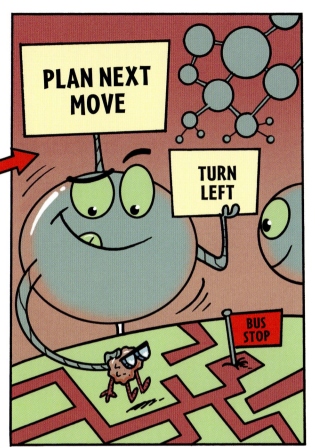

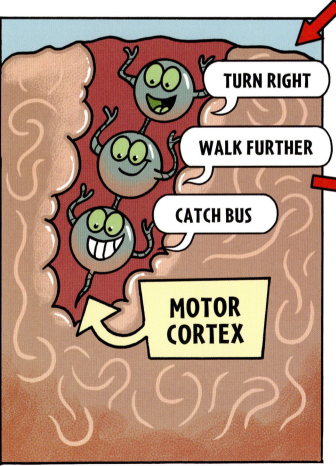

HOW PLANNING AHEAD WORKS

The brain chooses what to do now. But it also has to plan for later. How do you get to the shop? You may need to turn left, turn right, walk further, catch a bus ...

But if you did these things in the wrong order, you would not get there. A plan is a list of things to do, in the right order, to get what you want at the end. The brain has memory to keep plans, to remember the order, and what comes next.

STOP

To plan, we need to predict. Prediction means that we know what comes next. The brain is always predicting the future. If we can't predict where we will be, we cannot make a plan.

PUTTING THINGS IN THE BRAIN

Thinking lets us move ideas around in our head. Some ideas come from our senses. Sensing things outside us is called **perception**: we see, hear, touch, smell or taste. We also sense things inside us, like hunger and thirst.

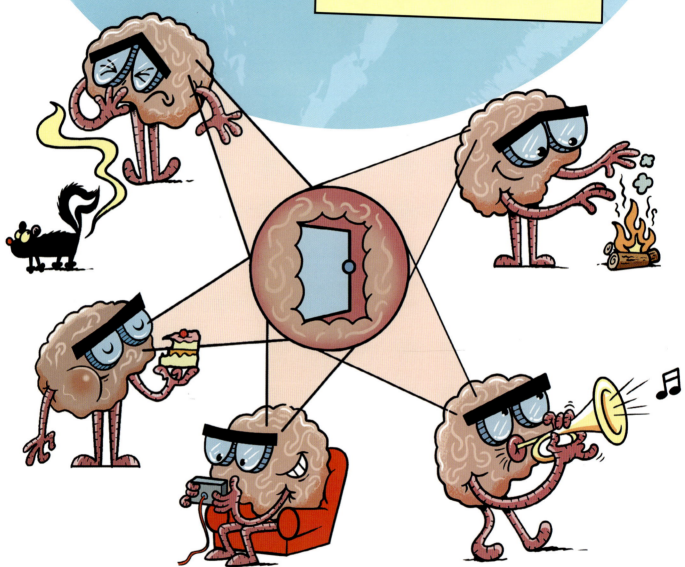

After we sense things, the information is stored in memory.

The brain can remember many things. Different brain areas remember different things:

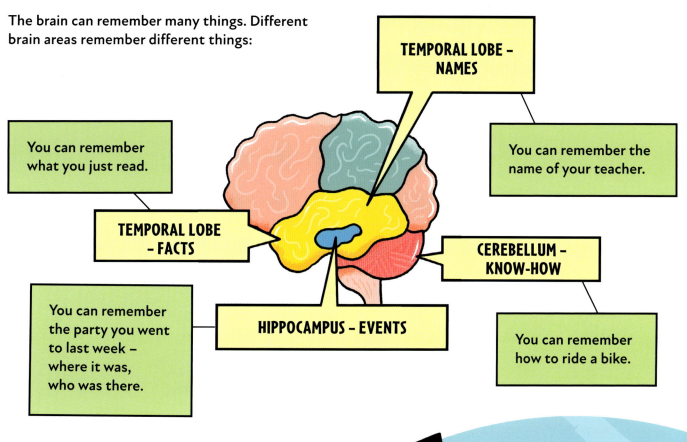

TEMPORAL LOBE – NAMES

You can remember what you just read.

TEMPORAL LOBE – FACTS

You can remember the name of your teacher.

CEREBELLUM – KNOW-HOW

You can remember the party you went to last week – where it was, who was there.

HIPPOCAMPUS – EVENTS

You can remember how to ride a bike.

The brain keeps all of these types of memory. Memory is like a notebook. You can write things in it, and find them later.

Some memories don't last very long – just seconds. Other memories will last your whole life.

MEMORY

BRAIN READS INSTRUCTIONS ON HOW TO BAKE A CAKE.

FLOUR

OVEN 210°
FLOUR 120G

FRONTAL LOBE

210

120

WORKING MEMORY

WAS IT OVEN 210 OR 120?

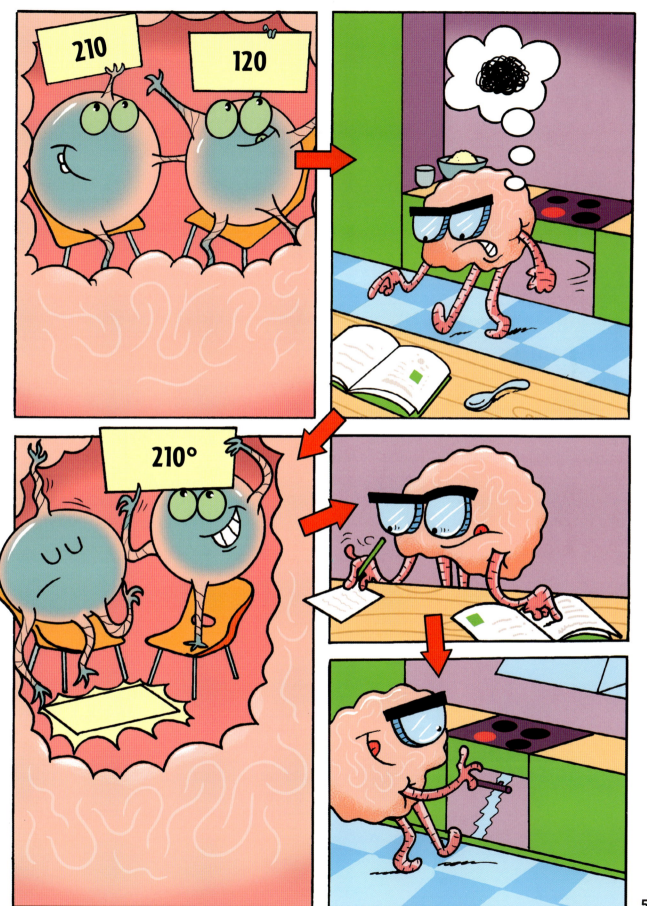

HOW MEMORY WORKS

We do thousands of things and we can still remember a lot of detail about what we did. This is because we store all the pieces of an event together.

An event is the place you were, the time it happened, the people you were with, how you felt, and what you did.

All these things are put together into a memory.

Old memories come back when something triggers them. This is called **retrieval**, or **recall**.

Strong feelings can make some memories very easy to recall. Sometimes, places and smells can bring them back.

WHO

WHEN

WHERE

The same brain areas that store memories can also tell us where we are, and can be used for thinking about the future.

People over fifty years old have fewer memory neurons, and so their memory gets worse. Some older people have to think harder to remember what has happened. **Dementia** is an illness that affects neurons and makes memory worse.

LEARNING SKILLS

BACK OF BRAIN

BRAIN GETS ON A SKATEBOARD.

TIPPING LEFT – PULL RED LEVER!

ERM ...

TOO LATE

WHOOPS!

ALWAYS WEAR A HELMET WHEN CYCLING OR SKATING. THE BRAIN IS SOFT AND CAN EASILY GET HURT.

HOW LEARNING WORKS

Learning is like memory, but takes time. When we practise things, we slowly learn more and more about what to do.
When we are babies, we have to learn to see things.

To see, the brain gets a pattern of light from the eye, and works out that it means there is something solid there.

When you see a shiny spoon, you know it is metal, and feels cold and hard.

Babies do not know this. They learn by playing with things.

Neurons gradually recognise shiny things and store how they feel. This is called **learning by experience**.

DID YOU KNOW?

Learning lets us change how we behave It can make hard things automatic. Some things are slow to learn and need lots of repeating. But we can learn facts very quickly.

Babies have to learn to stand. They don't know that pushing their legs helps them stand.

BABIES LEARN BY TRYING OUT LOTS OF THINGS.

SOMETIMES THEY WILL FALL, AND OTHER TIMES THEY WON'T.

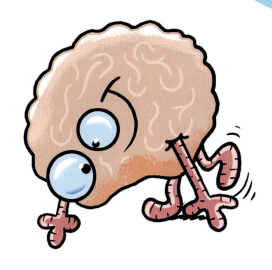

SLOWLY, THEY LEARN WHICH MUSCLES KEEP THEM UP. THIS IS CALLED 'MOTOR LEARNING'.

HEARING

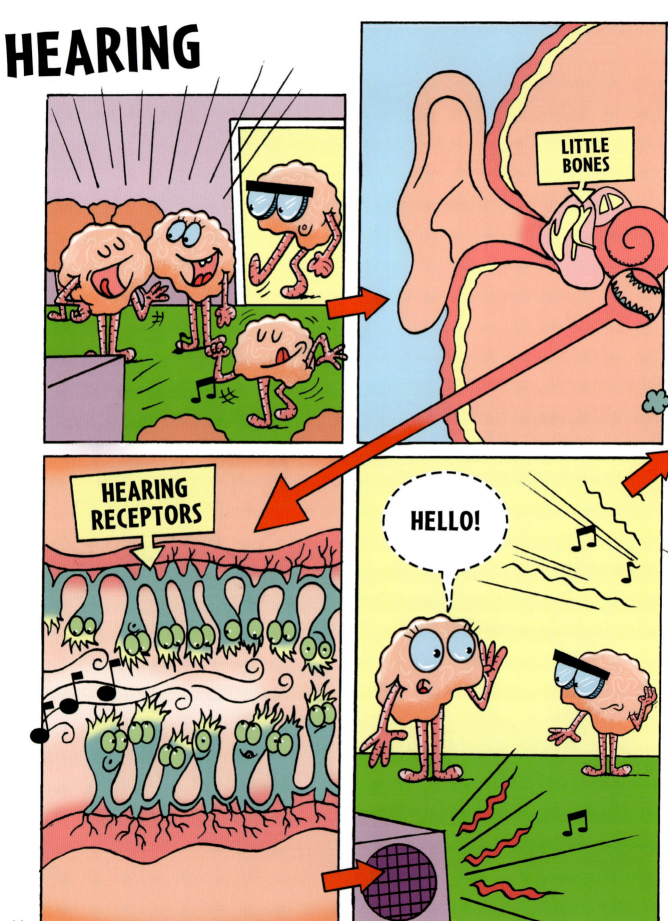

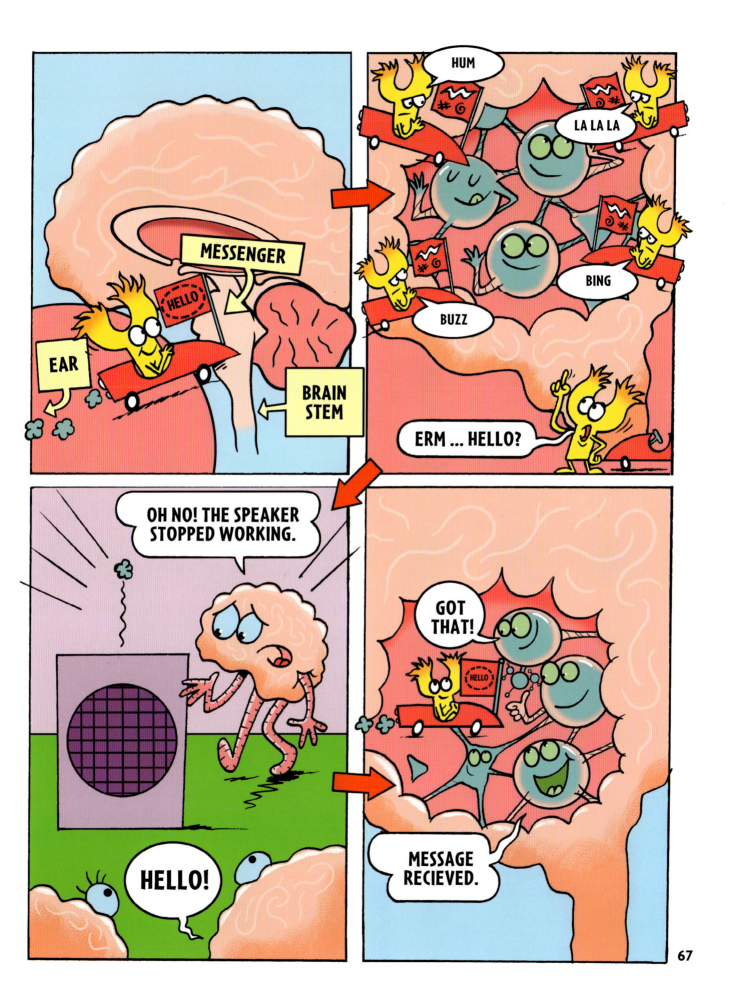

HOW HEARING WORKS

Sound is vibration – a very fast shaking. Sound shakes 20 to 20,000 times in one second. It makes cells in the ear wobble, and they send a message to the brain.

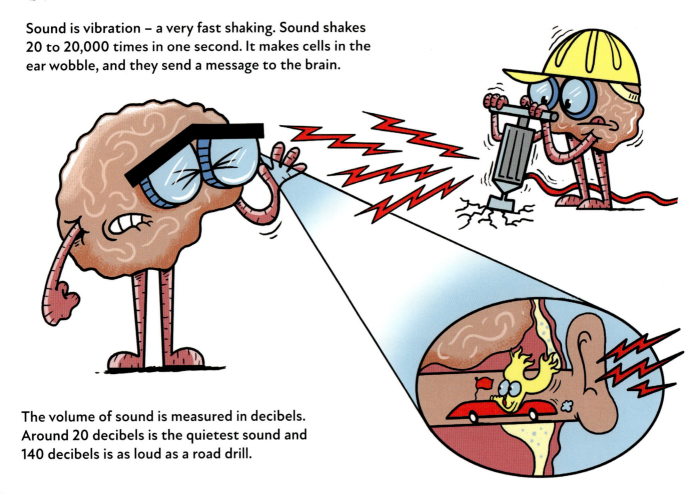

The volume of sound is measured in decibels. Around 20 decibels is the quietest sound and 140 decibels is as loud as a road drill.

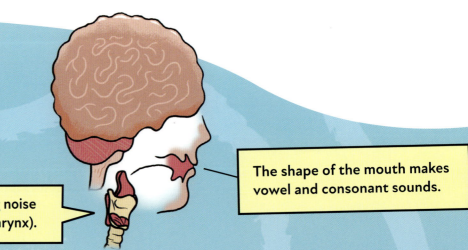

We talk by making noise in our voice box (larynx).

The shape of the mouth makes vowel and consonant sounds.

The brain puts words together, in the right order. This is called grammar.

In most people, the left side of the brain controls language. The left front area controls speaking and writing, and the left back area helps us understand both speech and written language. Damage to these areas makes speaking and understanding hard.

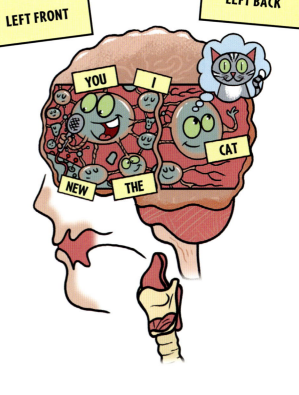

DID YOU KNOW?

Grammar has complex rules. Your brain obeys these rules without you even knowing!

COMMUNICATION

BRAIN PACKS A SCHOOL BAG.

MONDAY

TUESDAY

WEDNESDAY

MONDAY

TUESDAY

WEDNESDAY

WAIT ... WHAT?

WE NEED TO FIGURE OUT WHAT DAY IT IS.

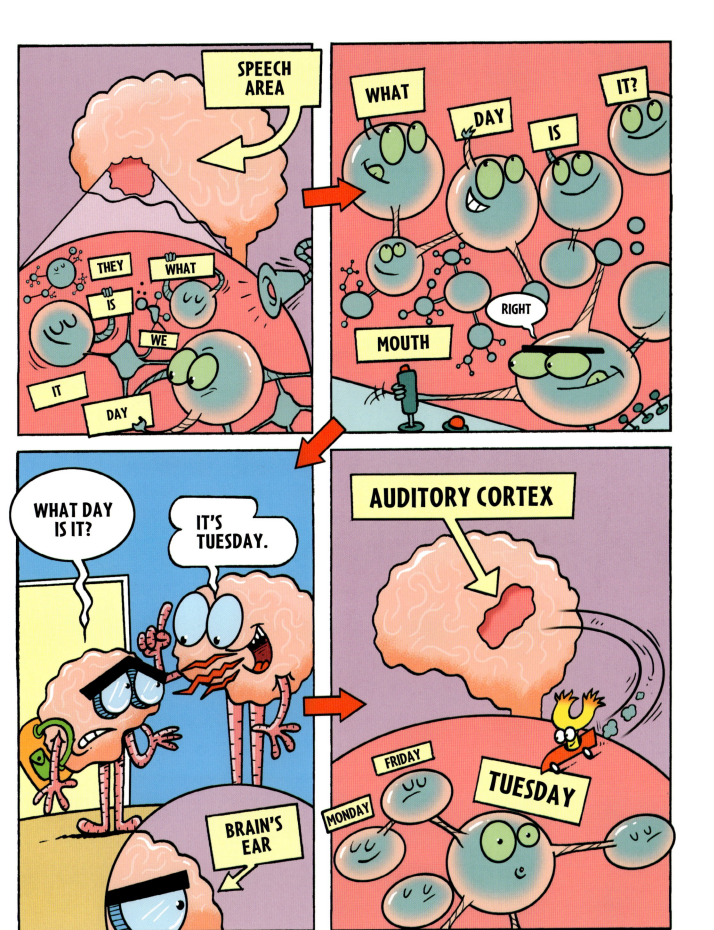

HOW COMMUNICATION WORKS

When someone answers a question, they give us **information**. Information is a signal that tells us something we didn't know. Information is the meaning of a message.

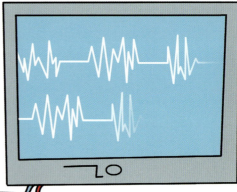

Communication is when one person gives information to another person. If you answer my question, information flows from your brain, to mine, as I hear you.

DID YOU KNOW?

Some animals have words or signs. But only humans join words into sentences.

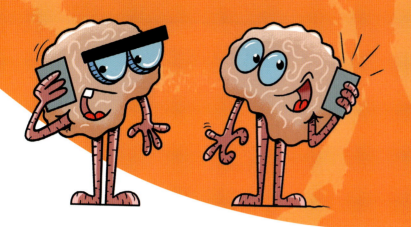

Information about how you feel goes from your brain to mine.

Neurons in the **auditory cortex** recognise sounds. They tell other brain areas which words you heard.

We communicate in many ways: for example, by phone, by talking, or by email.

We even communicate with our hands and faces. If you smile, it shows that you are happy.

NEURONS COMMUNICATE

Parts of the brain communicate with each other. When you think 'It is Tuesday', you first remember what day it is using memory. Memory neurons send this information, within the brain, to the speech area.

To get anything done, the parts of the brain have to send messages to each other.

73

BRAIN HAS AN IDEA!

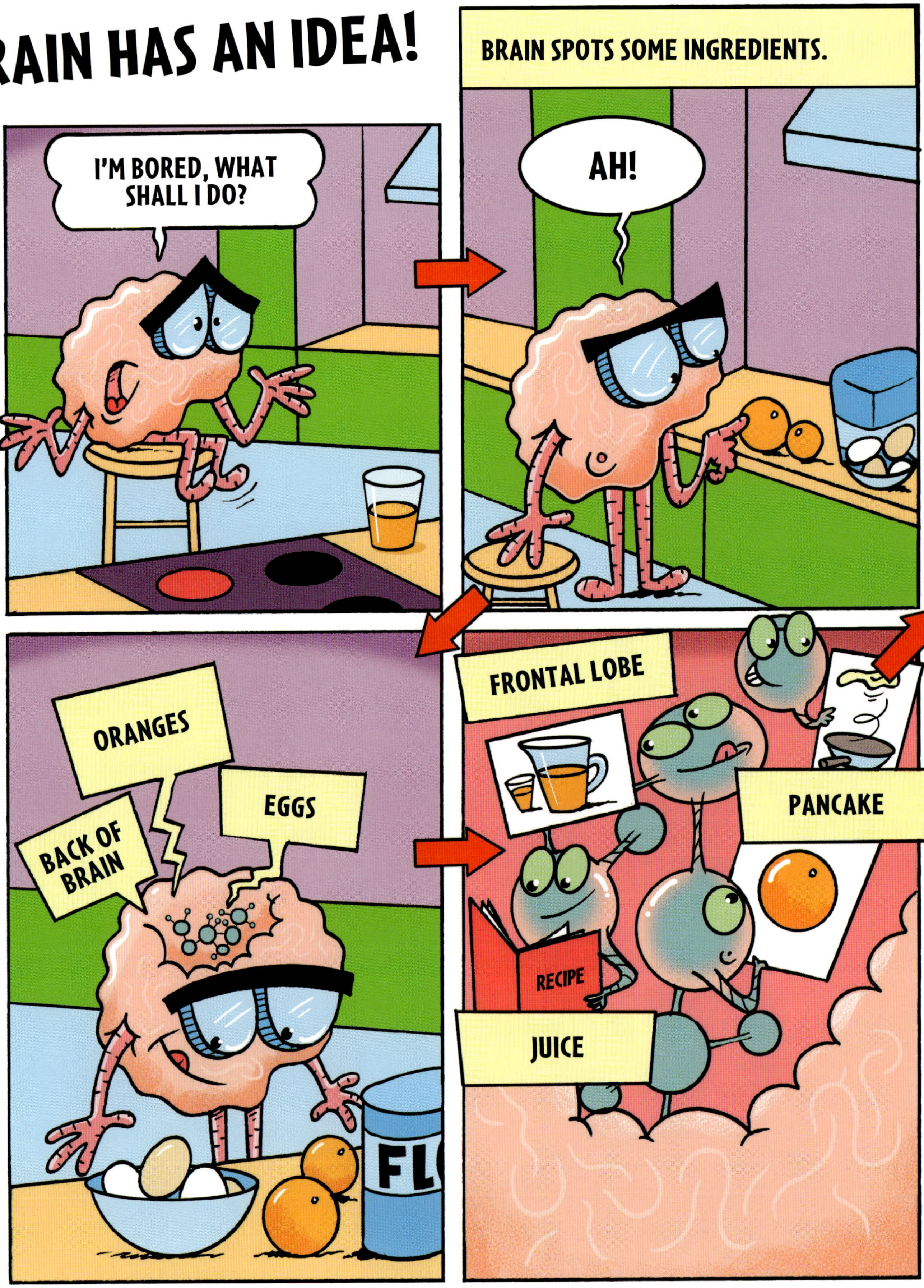

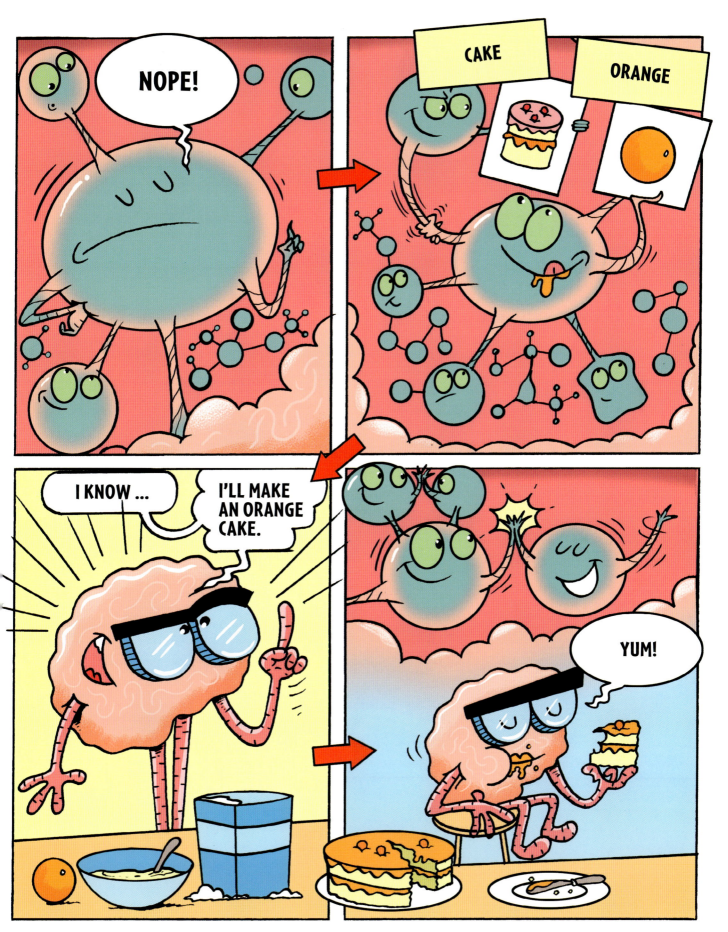

HOW IDEAS WORK

Humans beings are curious: we are always looking for information. You might ask "what is for dinner?". You want to know, before eating, what it will be. Curiosity helps us discover new things in the world.

When we get new information, it makes us believe things. If you smell food cooking, you may believe it is nearly dinner time.

If you hear the doorbell, you may believe somebody has come to visit.

When sensory parts of the brain get new information, your brain creates a new belief, or changes what it believes.

DID YOU KNOW?

The human brain can create new ideas. It takes bits of old ideas and mixes them up to make new ones. This is called **creativity**.

KNOW IT ALL

Knowing things is called knowledge. It is a kind of memory, stored all over the brain.

It is stored in the connections between neurons, called synapses. These connections make ideas and link them together.

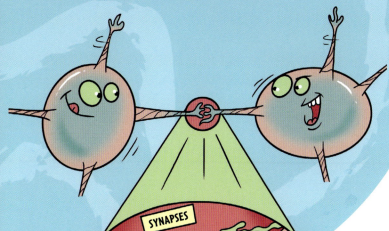

SYNAPSES

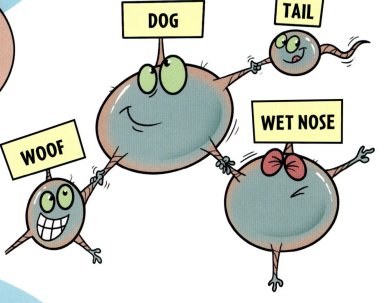

DOG

TAIL

WET NOSE

WOOF

Learning facts makes new connections between neurons. This lets us recognise things, and know their names.

The brain believes facts based on evidence – for example, from perception. But not everything we see or hear is true!

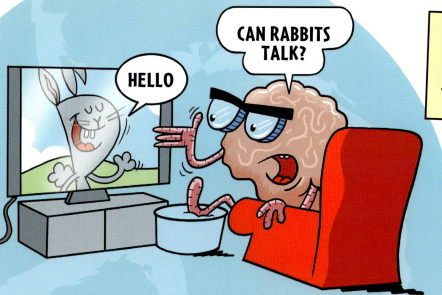

HELLO

CAN RABBITS TALK?

Before you believe anything, first decide whether it is really true. This is called **critical thinking**.

SWITCHING BETWEEN TASKS

We know lots of skills. But how do we know which skill to use?

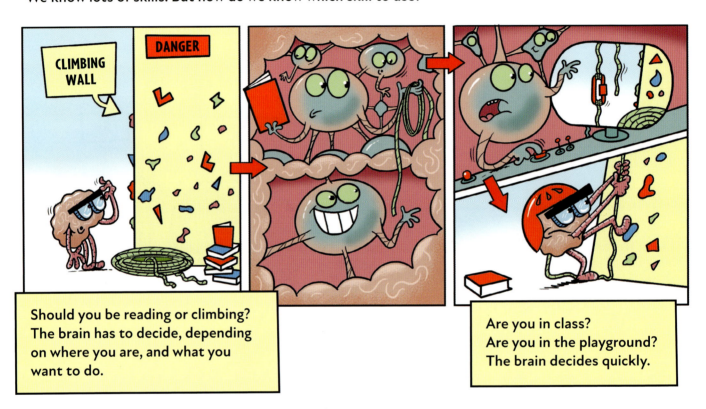

Should you be reading or climbing? The brain has to decide, depending on where you are, and what you want to do.

Are you in class?
Are you in the playground?
The brain decides quickly.

The sensory parts of the brain work out where you are. The frontal lobe keeps track of what you want, and chooses the right skills for the job. It tells the motor areas which skill to use.

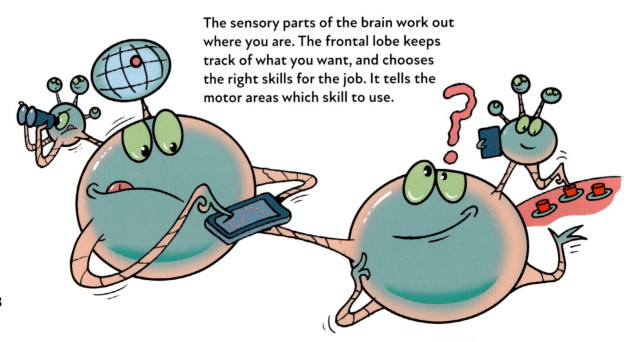

MAKING DECISIONS

DO I NEED A COAT OR AN UMBRELLA?

If the brain has two options, it has to choose. Neurons compete with each other. Each team **inhibits** the other. The winning neurons make the decision.

DID YOU KNOW?

Can you read two books at once? It is hard! What if you change books after every line? It's still hard! The brain takes time to switch over. Practice can make it much easier to switch from doing one thing to another.

A MAZE OF CONNECTIONS

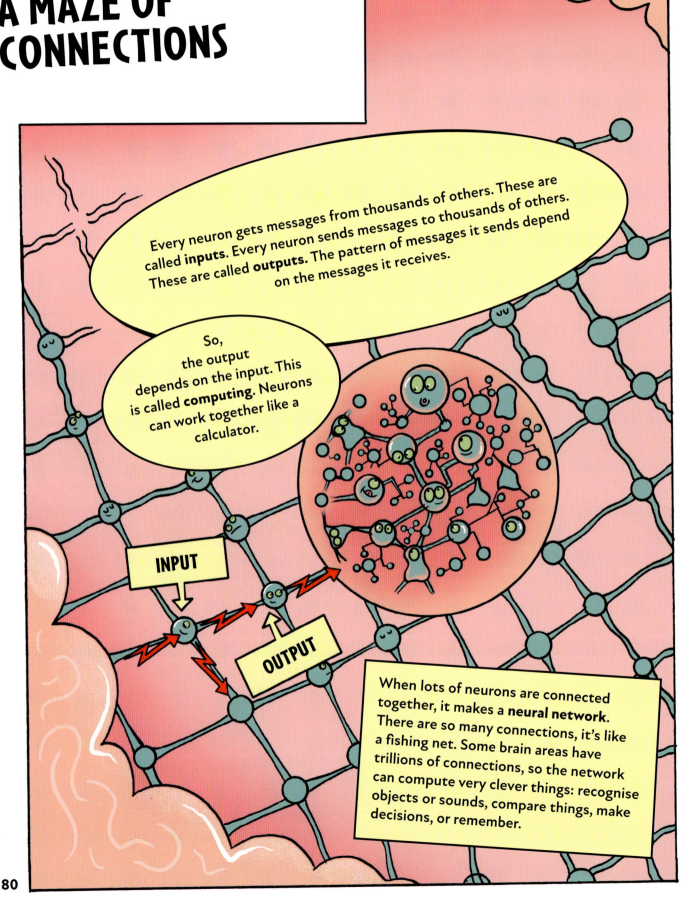

Every neuron gets messages from thousands of others. These are called **inputs**. Every neuron sends messages to thousands of others. These are called **outputs**. The pattern of messages it sends depend on the messages it receives.

So, the output depends on the input. This is called **computing**. Neurons can work together like a calculator.

INPUT

OUTPUT

When lots of neurons are connected together, it makes a **neural network**. There are so many connections, it's like a fishing net. Some brain areas have trillions of connections, so the network can compute very clever things: recognise objects or sounds, compare things, make decisions, or remember.

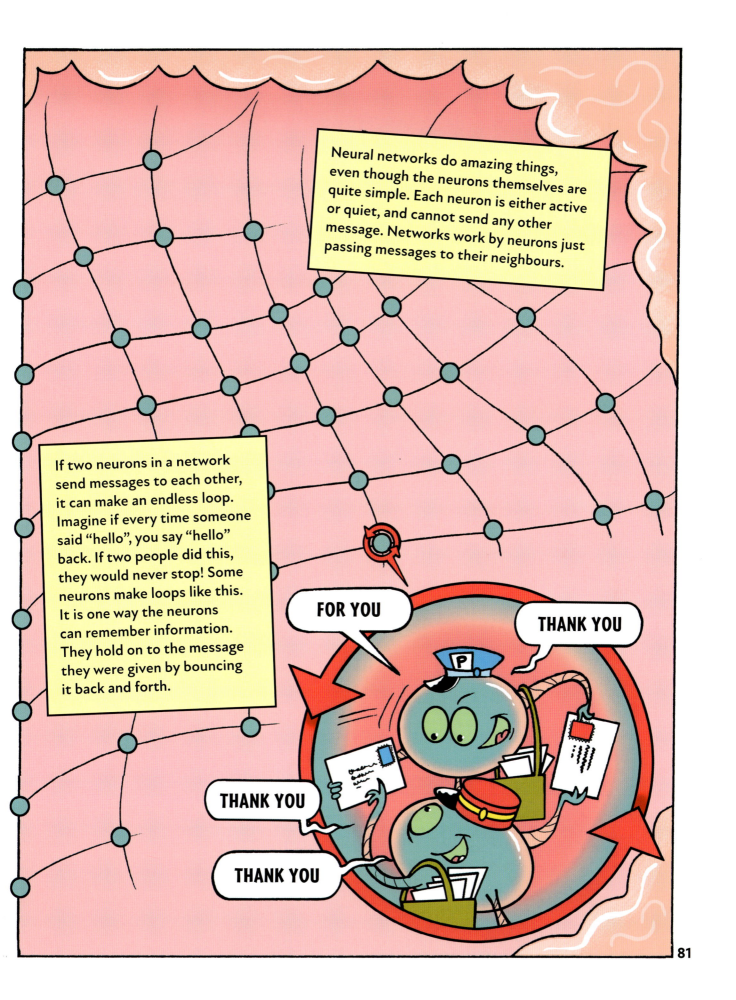

Neural networks do amazing things, even though the neurons themselves are quite simple. Each neuron is either active or quiet, and cannot send any other message. Networks work by neurons just passing messages to their neighbours.

If two neurons in a network send messages to each other, it can make an endless loop. Imagine if every time someone said "hello", you say "hello" back. If two people did this, they would never stop! Some neurons make loops like this. It is one way the neurons can remember information. They hold on to the message they were given by bouncing it back and forth.

FOR YOU

THANK YOU

THANK YOU

THANK YOU

WHAT ARE EMOTIONS?

Emotions make us feel good or bad.

SAD

They make us laugh and cry.

PROUD

They make us afraid or angry.

JEALOUS

They make us proud or jealous.

HAPPY

Emotions are strong feelings. As we grow older, we learn to control and use them.

Fear before a match or an exam can help us practice and improve. Anger when we fail makes us want to be stronger and better. These tricks are called **regulating** emotions.

We also learn to turn down the volume of bad emotions. This is called **inhibiting** them.

It is important to work out what caused the emotion, and to talk about it.

BAD EMOTIONS

DID YOU KNOW?

"Letting out" emotions can make us feel better. Crying, screaming or punching a pillow all let emotions out. But a better way is to use words.

EVOLUTION OF THE BRAIN

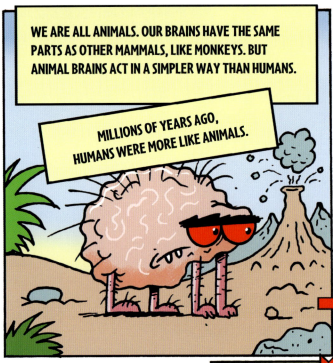

WE ARE ALL ANIMALS. OUR BRAINS HAVE THE SAME PARTS AS OTHER MAMMALS, LIKE MONKEYS. BUT ANIMAL BRAINS ACT IN A SIMPLER WAY THAN HUMANS.

MILLIONS OF YEARS AGO, HUMANS WERE MORE LIKE ANIMALS.

EVOLUTION MADE OUR BRAINS BIGGER AND MUCH CLEVERER.

THE BRAINSTEM IS THE DEEPEST, OLDEST PART. IT CAN CONTROL OUR RESPONSES.

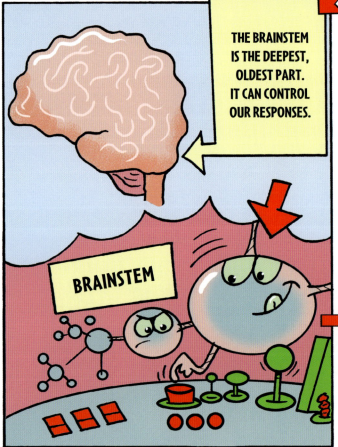

BRAINSTEM

BRAINSTEM RELEASES A MESSENGER INTO THE BLOOD CALLED ADRENALINE.

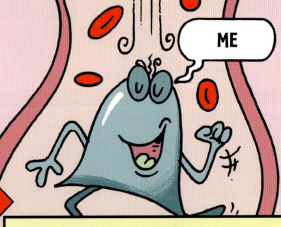

ME

IT TRIGGERS OUR HAIRS TO STAND UP. IT MAKES OUR SKIN COLD AND MAKES US SWEAT. IT STOPS OUR TUMMY FROM DIGESTING FOOD AND MAKES OUR HEART BEAT FAST, TO PUMP MORE BLOOD. STRONG EMOTIONS MAKE ADRENALINE, WHICH CAUSES THESE SIMPLE ANIMAL BEHAVIOURS.

HOW BRAIN EVOLUTION WORKS

The simplest brain needs only two neurons.

A slug's brain has a few more sensory neurons.

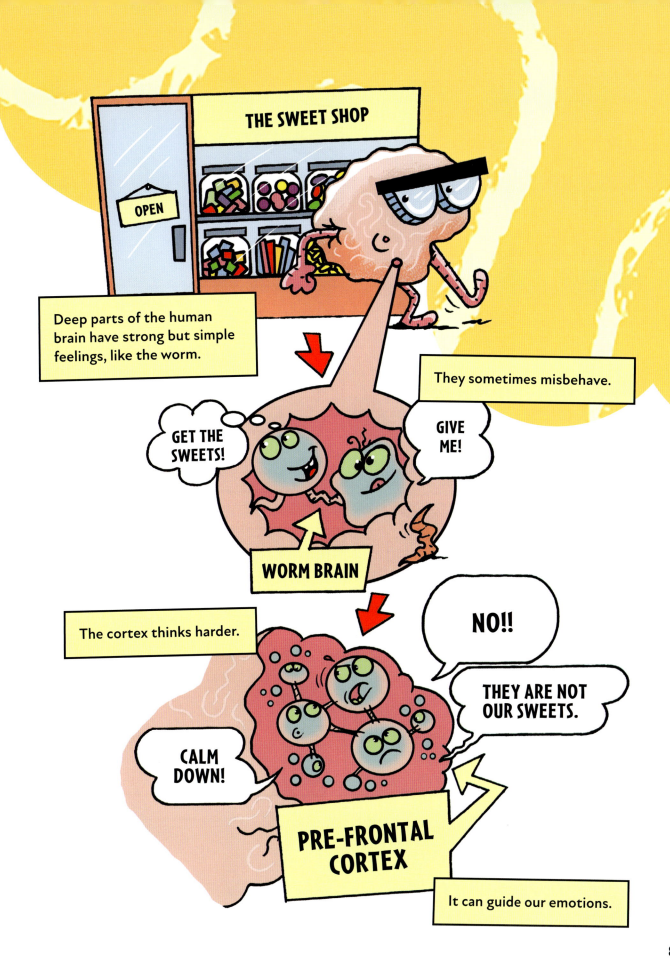

HUNGER AND THIRST

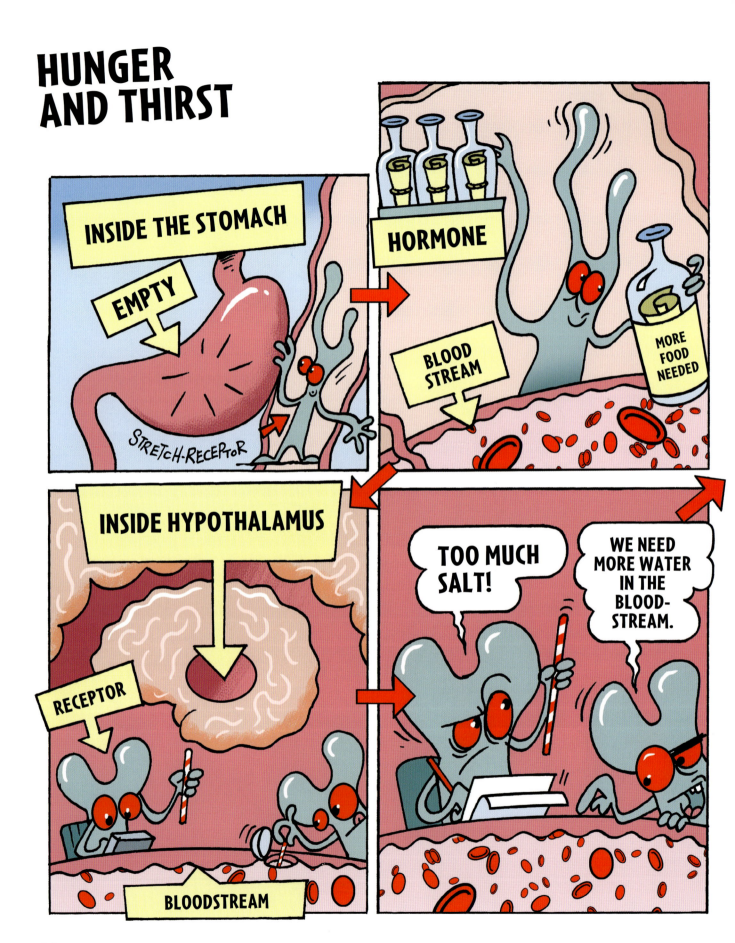

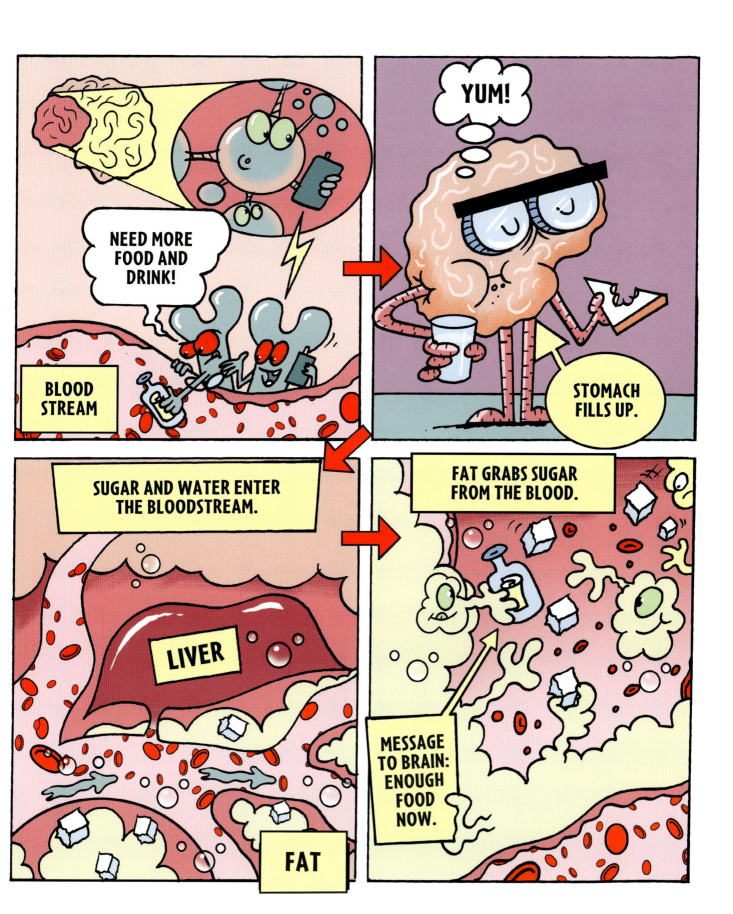

HOW HUNGER AND THIRST WORK

Hunger and thirst are drives. They make us want the things we need to survive.

The earliest animals had drives.

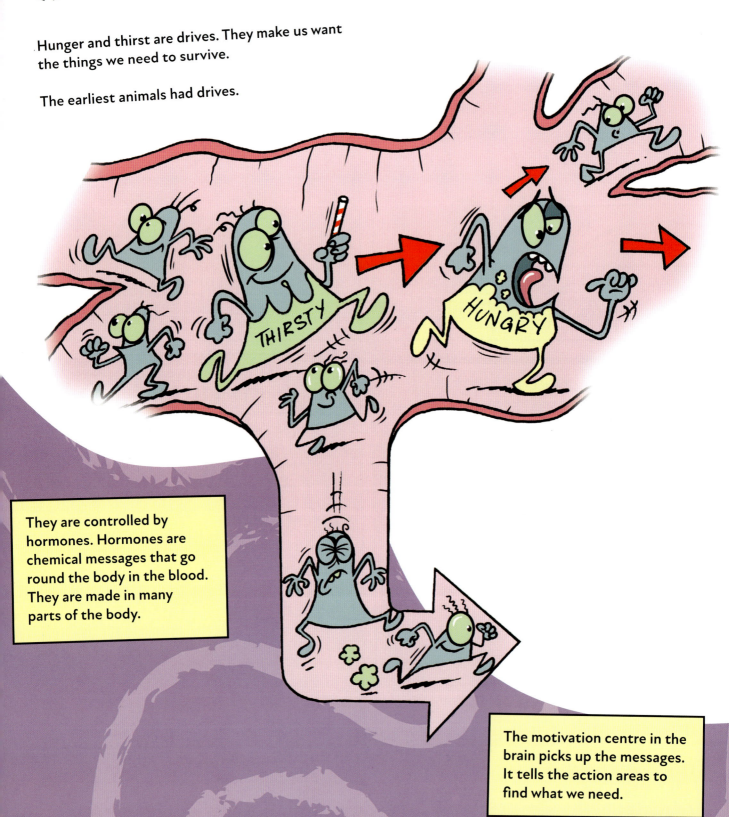

They are controlled by hormones. Hormones are chemical messages that go round the body in the blood. They are made in many parts of the body.

The motivation centre in the brain picks up the messages. It tells the action areas to find what we need.

THE DESIRE CENTRE

COMPLEX NEEDS

DID YOU KNOW?

The human brain can choose which goals to follow. It is very flexible. You could spend years and years getting good at a game! Choosing a goal, and working towards it, is called "motivation".

What happens if we don't get these needs?

If we don't get our needs on the top level we may feel lonely, disgusted, disappointed, worried, or guilty.

STRENGTH

FAMILY LOVE

MONEY HEALTH TOYS

AIR SLEEP z z z FOOD SHELTER

SIMPLE NEEDS

If we don't get our needs on the bottom level we get bad feelings like sickness, cold, fear, bitter taste, and itches. These feelings help protect us from danger.

Newer parts of the brain help us want more complex things. What sorts of things do you want?

FEAR

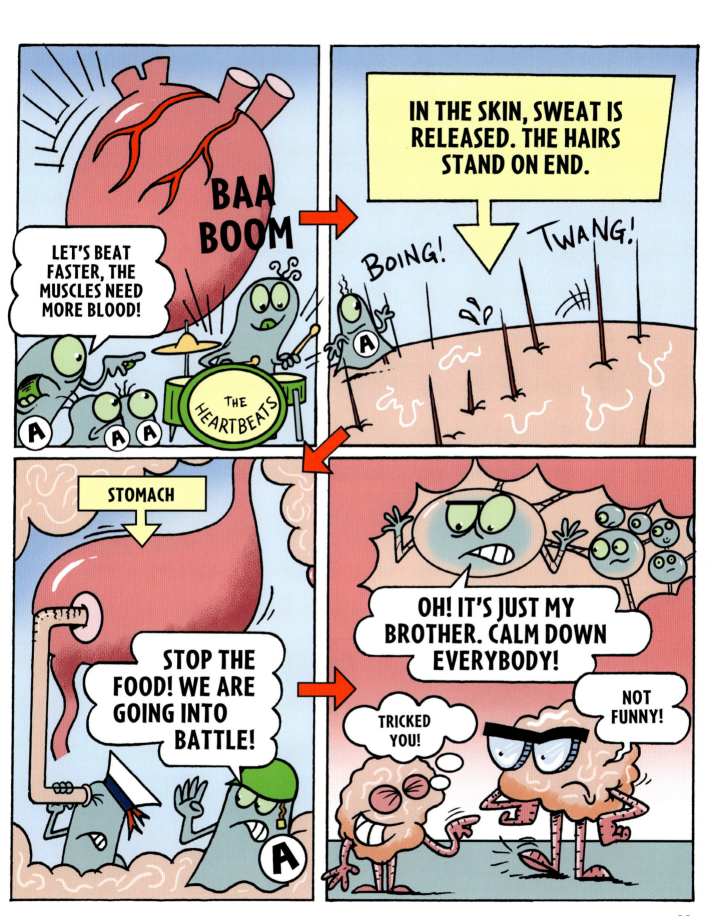

HOW FEAR WORKS

Adrenaline is a hormone.
Hormones are messengers sent in the blood.

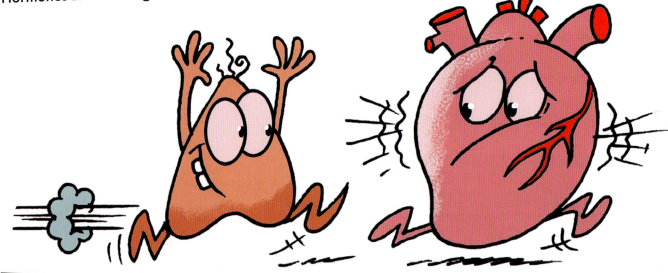

Adrenaline makes us ready for "fight or flight". We prepare to attack, or to run away. It is made when we are scared, excited or angry.

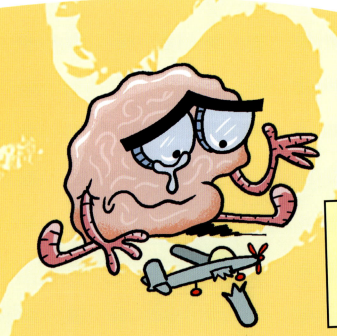

It usually happens when the old parts of the brain are fooled by something unexpected. Maybe your toy broke, or you lost a game. But instead, the old parts of the brain think you are in danger.

The "fight or flight" response lasts about five minutes.

– It makes your heart beat fast (heart racing or pounding).

– It makes you feel sick (butterflies in your stomach).

– You can have a cold sweat and panic.

Feeling all these things can make you even more worried.

THE CORTEX

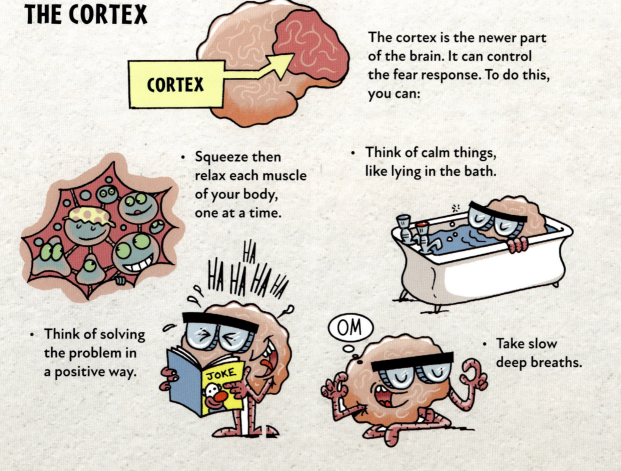

CORTEX

The cortex is the newer part of the brain. It can control the fear response. To do this, you can:

• Squeeze then relax each muscle of your body, one at a time.

• Think of calm things, like lying in the bath.

• Think of solving the problem in a positive way.

HA HA HA HA HA

JOKE

OM

• Take slow deep breaths.

These tricks help your cortex overcome adrenaline.

DIFFERENT MOODS

HOW DIFFERENT MOODS WORK

Mood is how we are feeling.
Mood can be positive or negative.

Positive moods are happy, interested, optimistic (thinking that something good will happen), or feeling content (feeling that you have everything you need).

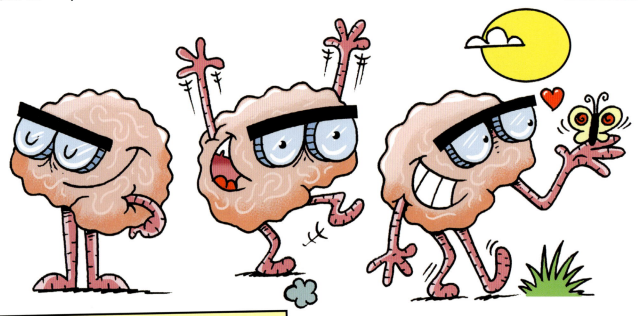

Negative moods feel sad, angry, worried, tired, irritable (feeling you will get upset or angry easily), and frustrated (feeling you are not getting what you want).

Moods can last a few minutes, and often last much longer. Moods change how we talk and behave.
Moods affect everyone around us.

When we feel happy, the deep brain areas make chemical signals. This helps us be more positive. Some people are born with more or less of these signals.
It is very hard to control your mood.
It can help to think carefully about why you are feeling bad.

CHEMICAL SIGNALS

If we think about what we should have done, we feel regret and guilt.

If we think of things that could have been better, we feel sad.

Almost everything has a good side, but we sometimes need practice to find it!

REACTIONS

HOW REACTIONS WORK

The brain is always thinking about other people. Other people have different ideas, and believe different things. Until about six years old, the brain can't imagine other people's thoughts very clearly. But after that, it can work out what someone else will think. The "social" brain develops.

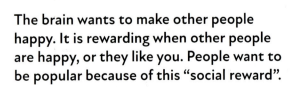

The brain wants to make other people happy. It is rewarding when other people are happy, or they like you. People want to be popular because of this "social reward".

We see minds everywhere. People see a fly, hovering around a doorway, and say "it is making up its mind".

If your computer is really slow, you might say "it is thinking", or "it's being silly".

We might see a pair of cars facing each other, and say they look happy together.

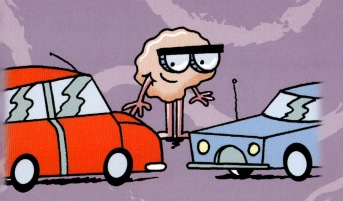

Probably the fly, computer, or cars, don't have a "mind". But we often talk like they do. It is normal and social to talk as if things are people.

SLEEP

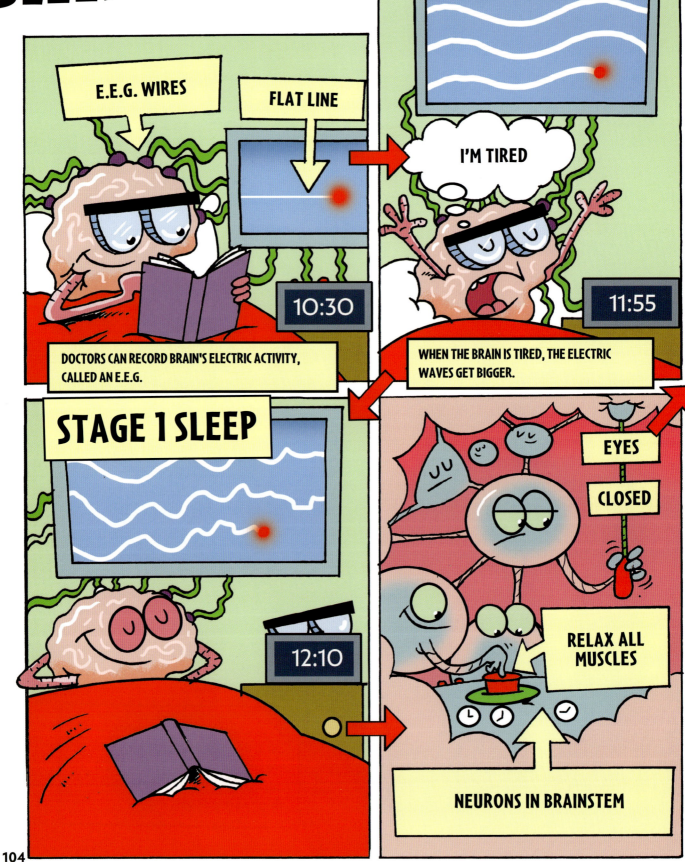

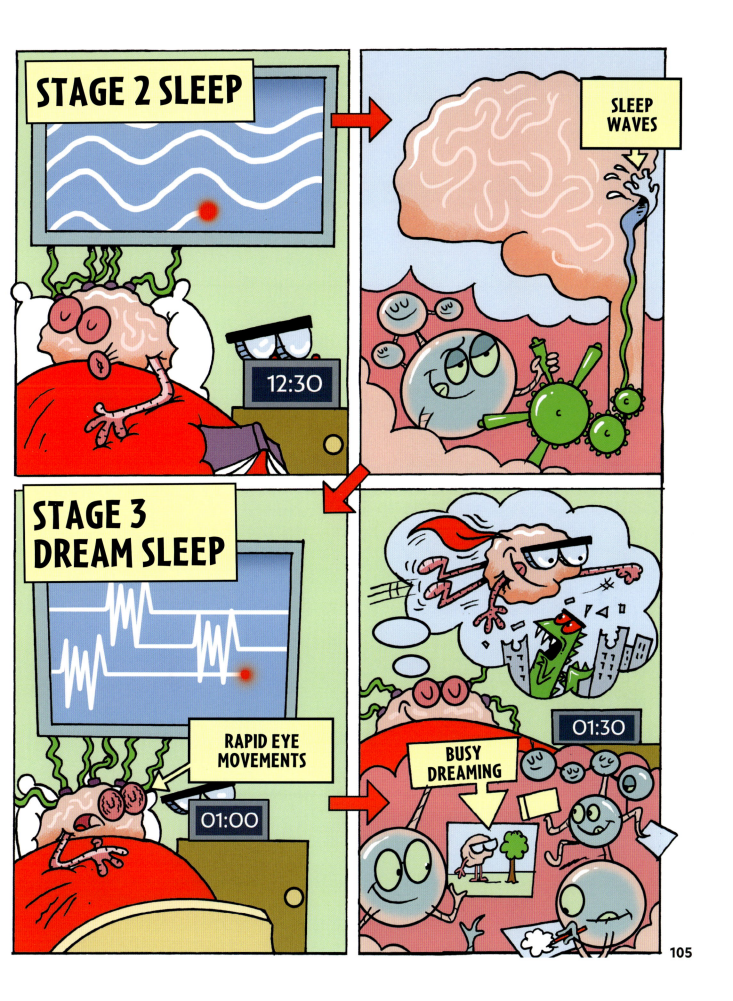

STAGE 2 SLEEP

SLEEP WAVES

12:30

STAGE 3 DREAM SLEEP

RAPID EYE MOVEMENTS

01:00

BUSY DREAMING

01:30

HOW SLEEP WORKS

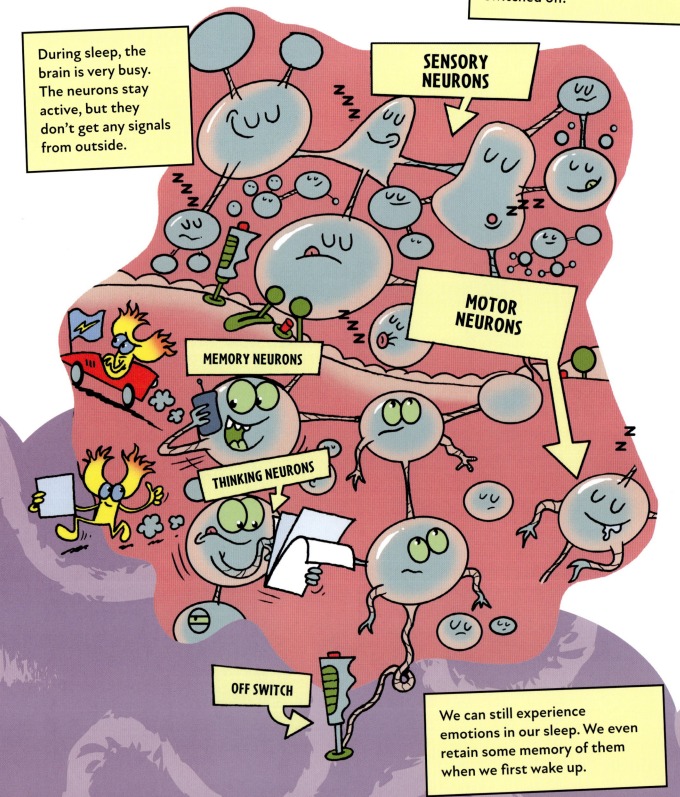

The sensory neurons are switched off. Also, our muscles don't move much. The motor neurons are switched off.

During sleep, the brain is very busy. The neurons stay active, but they don't get any signals from outside.

SENSORY NEURONS

MOTOR NEURONS

MEMORY NEURONS

THINKING NEURONS

OFF SWITCH

We can still experience emotions in our sleep. We even retain some memory of them when we first wake up.

When we dream, the thinking neurons in the cortex re-visit memories and imagine new things.

DID YOU KNOW?

Nearly all animal brains need sleep. Some animals sleep with one half of their brain at a time. The other half stays awake!

Some animals sleep but do not dream.

When the brain doesn't have enough blood, oxygen, or sugar, it switches off. This is called a coma. It is like sleep, but neurons are not active.

When we are conscious, we feel things. We know we are here. In dreamless sleep and coma, we are not conscious – we are unconscious. Scientists do not know what makes us conscious, yet.

In an operation, doctors can put people "to sleep". This is called a general anaesthetic. It makes the brain switch off, so it cannot feel anything.

THE USEFUL BRAIN

What we see, feel, believe, and decide happens in the brain. It's as if the brain has a model of the world inside it. It's not an actual model – the neurons calculate things about the world. Why is it useful to have the world in our head? It lets us make plans. When we need something, we can work out what to do.

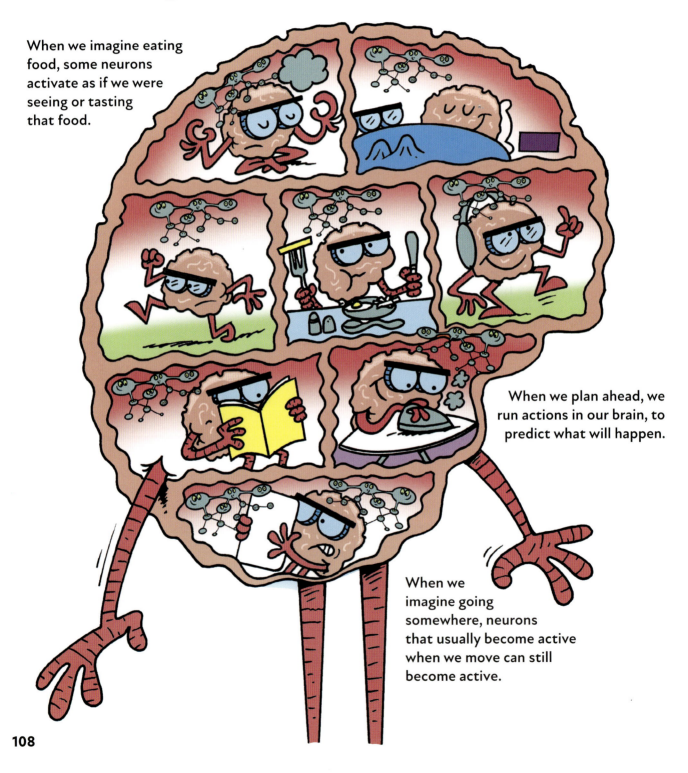

When we imagine eating food, some neurons activate as if we were seeing or tasting that food.

When we plan ahead, we run actions in our brain, to predict what will happen.

When we imagine going somewhere, neurons that usually become active when we move can still become active.

GLOSSARY

adrenals – glands at the top of the kidneys that make adrenaline

auditory cortex – brain area with neurons that notice sounds

axon – arms of a neuron that send messages out

brainstem – lower, stalk-like part of the brain, connecting to spinal cord

cerebellum – area at the back of the brain, important for coordination

communication – sending information from one place to another

compute – do sums that are needed to make choices

control – to be in charge of an action or behaviour

coordination – make parts of the body move in a controlled way

cortex – outer layer of brain's surface

creativity – having new ideas

critical thinking – to make a judgement after receiving information

decibels – the unit for measuring the loudness of sound

drives – reason for doing something

frontal lobe – the part of the brain that organises and controls other areas

hippocampus – brain area that remembers events and places

hormone – chemicals in the body that act as messengers

hypothalamus – area of the brain that makes hormones that control our temperature, heart rate, moods and hunger or thirst

information – signals that carry meanings

inhibit – to stop an action

input – the signal that goes into a neuron

nerves – fibres that carry information between the brain and the body

neural network – lots of neurons connected together to perform difficult tasks

neurons – brain cells that make simple decisions, receiving and sending messages

output – the signal that a neuron sends

perception – taking in sense information and experiencing it

posture – the position of the body

punishment – a negative reaction which helps us learn

receptor – cell that responds to senses

reflex – rapid response to something, without thinking

regulating – to control or adjust something

reward – a positive reaction which motivates learning

senses – hearing, seeing, touch, smell and taste

social reward – feeling good because of other people

spinal cord – bundle of nerves that run up and down your back

synapse – point where neurons touch to pass on messages

thalamus – centre of the brain, a relay centre for sensory and motor signals

vibration – moving back and forth. The air vibrates to make sound

INDEX

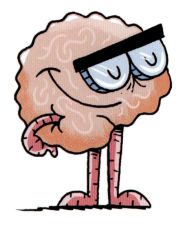

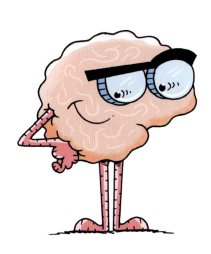

First published in Great Britain in 2024 by Wayland
Copyright © Hodder and Stoughton, 2024
Editor: Melanie Palmer
Illustration: Gary Boller
Interior design: Gary Boller and Jason Anscomb
Cover design: Jason Anscomb

ISBN: 978 1 5263 2333 0 HBK
ISBN: 978 1 5263 2332 3 PBK

Printed and bound in China
Wayland, an imprint of
Hachette Children's Group
Part of Hodder and Stoughton
Carmelite House
50 Victoria Embankment
London EC4Y 0DZ

An Hachette UK Company
www.hachette.co.uk
www.hachettechildrens.co.uk